Visions of STS

SUNY series in Science, Technology, and Society

Sal Restivo and Jennifer L. Croissant, editors

Visions of STS

Counterpoints in Science, Technology, and Society Studies

EDITED BY

Stephen H. Cutcliffe
and
Carl Mitcham

STATE UNIVERSITY OF NEW YORK PRESS

The cover photo is "View from Above, Grand Coulee Dam, 1998,"
by Stephen Cutcliffe.

Production by Diane Ganeles
Marketing by Michael Campochiaro

Library of Congress Cataloging-in-Publication Data

Visions of STS : counterpoints in science, technology, and society studies / Stephen H.
Cutcliffe, Carl Mitcham (editors).
    p. cm. — (SUNY series in science, technology, and society)
    Includes bibliographical references and index.
    ISBN 0-7914-4845-2 (alk. paper) — ISBN 0-7914-4846-0 (pbk. : alk. paper)
    1. Technology—Social aspects. 2. Science—Social aspects. I. Cutcliffe, Stephen H. II.
Mitcham, Carl. III. Series.

T14.5 .V574 2001
303.48′3—dc21

                                                                        00-038772

10 9 8 7 6 5 4 3 2 1

Contents

v

Part III. Critiques

Introduction:
The Visionary Challenges of STS

Understanding STS is a challenge—and in more than one way. First off, STS is not easily defined. Second, it challenges us to think about our scientific and technological society with greater depth than is often assumed to be necessary. Furthermore, STS is the outcome of more than one vision, and may be viewed from more than one perspective. To appreciate the rich and complementary character of these challenging visions, it helps to have some knowledge of the historical background out of which STS has emerged, and some preliminary profile of the spectrum of views collected in the present volume.

Historical Background and the Challenges of STS

The rise of modern science and technology has presented a series of special challenges to society. In the sixteenth and seventeenth centuries (with Galileo, Bacon, and Descartes) and again in the nineteenth century (with Darwin) conflicts arose between science and religion, none of which have ever abated. In the eighteenth and nineteenth centuries (with the industrial revolution) special problems arose for economics and politics, which neither socialism, capitalism, nor democracy have been able fully to resolve. The twentieth-century advent of nuclear weapons, electronic computers, and biotechnologies has only intensified these multiple challenges that range from issues of personal belief and social justice to nuclear risk, environmental pollution, cultural integrity, and self-identity.

1

The interdisciplinary field of STS is the most general attempt to map out and to assess such challenges, as well as the responses that have emerged. As such STS is itself a challenge to both routine acceptance of scientific and technological change and uninformed or narrow-minded reactions to such changes. STS takes up the challenge of the influence of science and technology on society. It becomes itself a challenge to uncritical acceptance of the world-historical transformation that began in the sixteenth century and has reached a dynamic crescendo as we enter the twenty-first.

The interdisciplinary STS challenge has roots in diverse disciplinary formations and cultural activities. The economic analyses of Adam Smith and Karl Marx, the novels of social consciousness by Dickens and Zola, and political reform movements that have gone under the names of liberalism, progressivism, and even neoconservatism have all played roles. But it is disciplinary specializations in the history and philosophy of science dating from the early part of the twentieth century, which were eventually followed by parallel disciplinary studies of technology and medicine, that have been the most salient influences in STS. During the mid-1960s various configurations of these scholarly pursuits—influenced as well by such activist initiatives as the consumer and environmental movements—became formally known as both the STS movement and STS studies (Cutcliffe 2000).

"STS" is actually a contested acronym. At first it stood for "Science, Technology, and Society"—and was characterized as a movement. Science, technology, and society programs emerged at various universities in the United States, Europe, and Australia, not always using this exact phrase. Examples include, for instance, the Science in a Social Context or SISCON program in the U.K. and the Values, Technology, Science, and Society or VTSS program at Stanford, both from the 1970s. When STS played a role in K-12 science education it was often time hyphenated as Science-Technology-Society and used as an adjective to qualify curriculum content. During the 1980s a number of university departments such as those at Cornell University and Rensselaer Polytechnic Institute reinterpreted the acronym to stand for science and technology studies, and took steps to transform the interdisciplinary field into a scholarly discipline with all the accoutrements thereof—from tenured faculty lines and degrees to journals and textbooks.

Early science, technology, and society programs often adopted as their representatives such figures as Jacques Ellul (1964) and Lewis Mumford (1967 and 1970). They presented global character-

izations of science and technology as independent or semiau-
tonomous forces dominating society, with at least implicit calls for
their active delimitation. Later science and technology studies
scholars have come to focus on the analysis and explication of specific sciences and technologies as complex societal influences and
social constructs entailing a host of political, ethical, and general
theoretical questions. In this "contextual" view, STS presents science and technology as neither wholly autonomous juggernauts nor
simply as neutral tools ready for any utilization whatsoever.
Instead, sciences and technologies are seen as value-laden social
processes taking place in specific contexts—interactively shaped by,
and in turn shaping, the human values reflected in cultural, political, and economic institutions.

Against this background of tensions between macro and micro
perspectives, STS challenges us to pursue interdisciplinary conceptualizations of the attendant complex interactions at both the individual and global levels. Medical science and technology—which, in
social constructivist analyses such as those by Bruno Latour and
Steve Woolgar (1979), merge into technoscience (Latour 1987)—obvi-
ously are influenced by and influence health care practices and policies. But they also may be linked with issues as seemingly remote as
stratospheric ozone depletion, since the presence of effective techno-
science treatments for skin cancer in Europe, North America, and
Australia make the developed world, which is the primary cause of
ozone depletion, more able to meliorate the consequences than
underdeveloped countries in South America, Africa, and Asia.

A second challenge of STS is, in both micro and macro analyses, to pursue interdisciplinarity by walking a fine line of judicious
analysis between promotional enthusiasm and oppositional rejec-
tion. In all its incarnations, despite repeated charges to the contrary,
it is crucial to note that STS is neither pro-science and technology—
what Langdon Winner has called HSTS, "Hooray for Science, Technology, and Society"—nor is it anti-science and technology. To call
even the STS movement anti-technology simply because it often
subjects science and technology to wholesale criticism is like calling
an art critic "anti-art" (Winner, 1986, p. xi; and 1989, p. 436).

A Spectrum of STS Visions

Given the contextual relationship between science, technology,
and society, and the generalized description of STS as a field of

study more than a discipline, it is natural that it exhibits many different approaches to issues and emphases. There is not just the challenges of STS; there are also many challenges in STS. Making sense out of the interdisciplinary complexities and the debates among different approaches to STS can be a daunting task. One aim of this book is thus to assist in such a sorting out process by providing a selection of brief statements representative of influential persons and perspectives in the broad STS field.

To this end we have invited ten STS scholars each to contribute short essays outlining their views on either the current state of STS or where the field may or should be headed. It is our hope, moreover, to enhance not just the understanding of science, technology, and society relationship but to advance intelligence in public decision-making with regard to science and technology.

To facilitate comparisons of the ten visions of STS we have divided them into three groups: general perspectives, applications, and critiques. This should be taken as more a heuristic classification than a rigid categorization, because in fact each essay makes some claim to advancing a general perspective, applying STS to particular problems, and criticizing inadequacies of the field. There are nevertheless differences in emphasis, and these are reflected in the provisional categorization that has been adopted.

Part I, "General Perspectives," includes four basic programmatic statements. Given the important role played by the question of technological determinism in STS discussions, it is appropriate to open with a chapter by Langdon Winner revisiting this issue. The truth is that although any comprehensive strong determinism has been widely rejected, it remains reasonable to argue a modified determinist thesis with regard to many aspects of technology. Indeed, as Winner effectively points out, there remains a recurring tendency in society at large to grant or adopt some kind of technological determinism.

Wiebe Bijker, the second essayist in this collection, makes a strong brief for what is called the social constructivist view of science, technology, and society. Social constructivism is the most systematically pursued program in the STS field; to some extent this view developed in opposition to and has largely replaced the research in technological determinism. Social constructivism has nevertheless been criticized as sometimes coming close to adopting a promotional or apologetic stance toward science and technology. Bijker restates the social constructivist stance as an attempt precisely to steer the challenging course between the barking Scylla of determinism and the swallowing Charybdis of endorsement.

In the third chapter, Lars Fuglsang, with a sketch of three general approaches to STS in relation to public policy formation, considers two versions of determinism: one in which science and technology shape society, another in which society shapes science and technology. As a more sound basis for policy formation and public action, Fuglsang argues for the interactionist approach that has become somewhat characteristic of the STS field.

Susan Cozzens, in a fourth essay, considers the problem of interdisciplinarity, especially from an academic standpoint. For her, there is also a problem of determinism, but by the inherited disciplines that contribute to any general STS understanding. The general challenge in STS is to work at transcending these multiple disciplinary divides.

The next three chapters, which constitute Part II, "Applications," all choose to stress more specialized perspectives in and on STS. Rudi Volti offers an STS perspective on technology and work, challenging some traditional views. Robert E. Yager, in a further challenge to college-based STS discussions, describes the role STS plays in educational theory, especially as influential in primary and secondary schools. Albert H. Teich examines STS from the perspective of a policy analyst. Each of these three applied visions uses STS not only to raise questions about popular assumptions regarding science and technology in contemporary society, but also to envision new ways of doing STS itself.

The need to envision new ways of doing STS, and thereby to renew the field, becomes the major theme of Part III, "Critiques." Richard Sclove takes an ironic approach, pointing out that on "other planets," such as some European countries, STS involves much more than just studying science and technology issues; it involves as well doing something about them. A robust STS attitude requires bridging the theory-practice divide.

Eulalia Pérez Sedeño subjects STS to a feminist criticism. For her the greatest weakness in STS has been the failure to appreciate the masculine biases of much science and technology, and the ways in which science and technology have differentially impacted women and men. Her argument may well have related implications for ways STS has failed to appreciate the differential impacts on various ethnic groups.

The collection concludes with Wilhelm E. Fudpucker's challenge to STS to rethink itself in light of fundamental transformations taking place in technology. Too often, he implies, and too easily, STS has assumed that it knew what the science or the technology is with which society interacts. It is not just conceptions of the theory-practice divide or disciplinarity or feminism that constitutes a challenge to STS; it may even be our conception of technology.

As indicated, these essays include several general assessments of the field as a whole (Winner, Bijker, Fuglsang, and Cozzens). There are also specific calls for more effective democratic participation in science and technology decision-making (Sclove) in the face of a concern regarding the deterministic nature of technology (Winner). Three pieces focus on applied themes in STS (Volti on work, Yager on education, Teich on science and technology policy). Others critique the effectiveness of STS as currently constituted (Pérez Sedeño and Fudpucker).

These diverse visions—appropriate for a truly interdisciplinary field—are representative of a diversity of authors from a diversity of contexts. The authors come from universities and independent institutes or professional organizations, as well as from as many as five different countries. Both younger and more well established practitioners of STS are granted an opportunity to present their visions and challenges. Taken together, these essays thus offer an exciting overview of the STS field, one that provides readers a kaleidoscopic perspective on many science and technology issues.

Collective and Independent Uses

The order of presentation may not always be the order that a reader may want to make use of these essays. In fact, each essay stands alone, and may be fruitful in any number of combinations.

Each of the essays is preceded by a brief headnote that summarizes its main theme, indicates the author's background, and raises one or more questions to consider while reading the essay. The aim here is to emphasize our effort to provide, not a single authoritative interpretation, but rather a series of ideas on STS approaches that will allow students and general readers to better grasp science and technology issues and as a result to exercise more informed citizenship with regard to science and technology in their lives. It is our hope thereby to enhance not just understanding of the relationships between science, technology, and society, but public decision-making with regard to science and technology as well.

Acknowledgments

Thanks are due to Professor Franz Foltz (Rochester Institute of Technology) for suggesting the idea of this collection. We also want to acknowledge the secretarial and research assistance of Karen Snare, Mark Pitterle, James Frazier, and, most especially, Abby Hoats.

S.H.C.
C.M.

References

Cutcliffe, Stephen. 2000. *Ideas, Machines, and Values: An Introduction to Science, Technology, and Society Studies*. Lanham, MD: Rowman and Littlefield.

Ellul, Jacques. 1964. *The Technological Society*. Trans. John Wilkinson. New York: Knopf.

Latour, Bruno. 1987. *Science in Action: How to Follow Scientists and Engineers through Society*. Cambridge, MA: Harvard University Press.

Latour, Bruno, and Woolgar, Steve. 1979. *Laboratory Life: The Social Construction of Scientific Facts*. Beverly Hills, CA: Sage.

Mumford, Lewis. 1967 and 1970. *The Myth of the Machine*. Vol. 1: *Technics and Human Development*. Vol. 2: *The Pentagon of Power*. New York: Harcourt Brace Jovanovich.

Winner, Langdon. 1986. *The Whale and the Reactor*. Chicago: University of Chicago Press.

———. 1989. "Conflicting Interests in Science and Technology Studies: Some Personal Reflections." *Technology in Society*, vol. 11, pp. 433–38.

I

General Perspectives

1

Where Technological Determinism Went

LANGDON WINNER

One of the central debates that has animated the STS field has concerned whether and to what extent one can accurately describe technology as constructing society or society as constructing technology. The former thesis is associated with the idea of technological determinism and notions of technology as a semi-autonomous force dominating other basic social institutions. The latter is associated with a research program that often goes under the name of the social construction of technology.

Langdon Winner's first book, *Autonomous Technology: Technics-out-of-Control as a Theme in Political Thought* (1977), established him as a leading defender of some aspects of the technological determinism thesis. His second book, *The Whale and the Reactor: A Search for Limits in an Age of High Technology* (1986), advanced his argument, especially in the chapter "Do Artifacts Have Politics?" Here Winner offers a brief restatement of one aspect of his thesis by noting the ironic way in which many of those who are the strongest defenders of technology, as fundamentally beneficial, seem at the same time to assume that it is impossible to alter the direction of its development. Interestingly enough, another irony is that Winner is professor of science and technology studies at an engineering school, Rensselaer Polytechnic Institute.

In reading this essay we may want to ask ourselves about our own assumptions with regard to the influence of technology on society. Do we not tend to assume that human history has been defined by its technologies, for instance when we speak of the "stone age," "iron age," or the "computer age"—and that all technological change is for the good? Do we think technologies should just be accepted and adjusted to? Have we ever tried to alter a technology and found it rather difficult to do so? It is precisely such questions that can help us assess the strength (or weakness) of the idea that technology may well exhibit a certain autonomous force—especially in our culture. From Winner's perspective, it may well be that technology is socially constructed—but that we have so socially constructed technology as to make technology the fundamental constructive influence in our world.

Among scholars in technology studies during the past quarter century there has been an enormous effort to discredit notions of determinism. As a result, within contemporary scholarly communities, once popular discussions of technological inevitability, determinism, and imperative have gone out of fashion. A few decades ago, debates about technology and society often focused on what were widely (but by no means uniformly) believed to be the essential features of technology and technological change. Although the general technology-using public did not think so, many economists, historians, and social theorists argued that the development and use of technology followed a fairly linear path, that technological change was a kind of univocal determining force with a momentum and highly predictable outcomes.

There were optimistic and pessimistic versions of this notion. Among social scientists, one influential group espoused what was called "modernization theory," the belief that all societies move through stages of growth or development linked to technological sophistication. Eventually they reach what one theorist called the "take-off point," thereby achieving the kind of material prosperity and way of life found in late-twentieth century Europe and America (see, for example, Rostow, 1960). There were also pessimistic variants of this conception, theories of technological society that focused on the human and environmental costs of rapid technological development, as in the critical visions presented by Jacques Ellul's *The Technological Society* (1964), Herbert Marcuse's *One-Dimensional Man* (1964), and Lewis Mumford's *Myth of the Machine: The Pentagon of Power* (1970).

Whether taken in optimistic or pessimistic variants, there was something of an agreement that modern technology had certain essential qualities, among which one could list a particular kind of rationality—instrumental rationality, the relentless search for efficiency—and a kind of historical momentum that rendered other kinds of social and cultural influences on the character of social life far less potent. Such views were largely in opposition to the more popular view of science and technology as a neutral phenomena that could be adopted by a variety of societies.

More recently, however, scholars have attacked from every angle the idea that modern technology is a univocal, unilinear, and self-augmenting force. Many scholars use seemingly cogent demonstrations to argue that technological devices, systems, and methods are socially shaped and thoroughly contingent products of human interaction. Technology has been interpreted as always subject to complicated "social shaping" or "social construction." Looking closely at how technologies arise and how they are affected by the contexts that contain them, one does not find a juggernaut foreordained to achieve a particular shape and to have particular consequences, but rather a set of options open to choice and a variety of contests over which choices will be made.[1]

Debunking work of this kind has been undertaken by European and American social scientists, historians, and philosophers. One purpose of this work is simply to provide a more faithful account of how technological innovation and associated social change actually occurs. Another goal is to snatch human choice from the jaws of necessity, to redeem the technological prospect from both the facile optimism of liberal, enlightenment thought and the pessimism of cultural critics. Hence, an endless array of case studies and social theories now proudly affirm voluntarism in technological change in contrast to notions of determinism.

As a way of thinking about past developments (or even of imagining possibilities in our own time) the new approaches have much to recommend them. But it is ironic that at the very moment that notions of contingency and social construction of technology have triumphed among social scientists and philosophers of technology, in the world at large it seems increasingly clear that unstoppable, strongly deterministic, technology-centered processes rule our times. In the literature of a wide variety of technical fields, the language of momentum, trajectory, technical imperatives, and determinism is more insistent now than in the naive 1950s.

The perception that we are swept along by a law-driven process of technological change is, for example, widely embraced among those who work in fields of computers and telecommunications. Thus, one of the founders of Intel, Gordon Moore, is widely acclaimed for having formulated "Moore's Law," which states that the computing power available on a microchip doubles roughly every eighteen months—forcing computer users into constant upgrades. Writers on computing and society have seized upon this "law" to account for the common perception that social change is driven by necessities that emerge from the development of new electronic technology as from nowhere else. As Stewart Brand (1995) explains to the readers of *Wired* magazine, "Technology is rapidly accelerating and you have to keep up."

Among many economists, businesspersons, and politicians there is also an openly deterministic vision of technological change seemingly oblivious to the new vision of historically contingent, socially constructed, and endlessly negotiable technical options. For example, in Lester Thurow's book *The Future of Capitalism* (1996), we learn that technological change is one of the "tectonic forces" that we can only obey but never hope to master. The popular on-line essay, "Cyberspace and the American Dream: Magna Carta for the Knowledge Age," written by Esther Dyson, Alvin Toffler, and others, describes the dynamic thrust of the digital revolution as our true destiny. "As it emerges," the manifesto explains, "it shapes new codes of behavior that move each organism and institution—family, neighborhood, church group, company, government, nation—inexorably beyond standardization and centralization."

Across diverse fields there is a strong tendency to conclude that rapid changes in technology and associated developments in social practice can only be described by a reformulated evolutionary theory, a theory of biotechnical evolution. Notions of this kind inform the speculations of the Santa Fe Institute about the emergent properties of complex biological and artificial systems. Summarizing implications of this way of thinking and applying it to contemporary development in the spread of networked computing, Kevin Kelly, the editor of *Wired* magazine, concludes:

> We should not be surprised that life, having subjugated the bulk of inert matter on Earth, would go on to subjugate technology, and bring it also under its reign of constant evolution, perpetual novelty, and an agenda out of our control. Even without the control we must surrender, a neobiological technology is far more rewarding than a world of clocks, gears, and predictable simplicity (Kelly, 1994, p. 472).

While new theories about the social construction of technology emphasize multiple sources of innovation, numerous branching points, and continuing negotiation among social groups, a widely shared conviction about technological change held just about everywhere else is something quite different. It holds that people must scramble to catch up with developments whose course is, for all intents and purposes, beyond deliberation and judgment. "Negotiation, social construction, and choice?" Get real! From the president of the United States on down, people are inclined to describe the future as one dominated by the forces of computerization, globalization of production, and other insistent technology-rooted trends. The smart people are those able to "re-engineer" their organizations and careers by liquidating older roles, relationships, and institutions in response to technical and economic necessities that loom ahead. The less proactive to these conditions are doomed to suffer as the new technical order crashes in on them.

The mood today—in its frantic rush to transform the schools in the image of Internet, for example—makes the debates about technology policy as recent as the 1970s seem democratic and voluntaristic in the extreme. Then at least there was a sense that social choice about alternative technologies—in energy, for example—was something worth debating in public. The abolition of the Office of Technology Assessment in 1995 is merely another sign that, at present, the range of options and negotiations open to most people is narrowing, not expanding.

I find it interesting that the scholarly community in STS is so inward looking that it seems not to notice the glaring disconnect between its own favored theories and the visions of run-away technology that prevail in society at large. True, the new methods and models are useful for historical study—reconstructing choices that have already been made, speculating about how outcomes might have been different, and so forth. But as for ways to illuminate matters before us right now, the new notions (contingency, interpretive flexibility, actor networks, and the like) seem distinctly impotent.

Perhaps the exposition of a wealth of choices supposedly available to us was merely an academic exercise anyway. Far from embracing the promise of humane, voluntaristic, self-conscious, democratic, social choice-making in and around technology, a great many observers have—for reasons they find compelling and completely congruent with their lived experience—cast their lots with ideas that reject or even mock choice-making of that kind.[2]

Notes

1. See, for example, Wiebe Bijker et al., eds. (1987). For my critique of this way of thinking, see "Social Constructivism: Opening the Black Box and Finding It Empty" (1993).

2. This article draws heavily on two previous essays: my "Technological Determinism: Alive and Kicking?" (1997) and my "Technology Today: Utopia or Dystopia?" (1998).

References

Bijker, Wiebe E., Hughes, Thomas P., and Pinch, Trevor, eds. 1987. *The Social Construction of Technological Systems: New Directions in the Sociology and History of Technology.* Cambridge, MA: MIT Press.

Brand, Stewart. 1995. "Two Questions," in "Scenarios: The Future of the Future." *Wired,* vol. 3, no. 11 (October), pp. 28–46.

Dyson, Esther, Gilder, George, Keyworth, George, and Toffler, Alvin. 1994. "Cyberspace and the American Dream: A Magna Carta for the Knowledge Age." Release 1.2. Washington, DC: Progress and Freedom Foundation, August 22, 1994. At http://www.townhall.com/pff/position.html.

Ellul, Jacques. 1964. *The Technological Society.* Trans. John Wilkinson. New York: Knopf.

Kelly, Kevin. 1994. *Out of Control: The Rise of Neo-Biological Civilization.* Reading, MA: Addison-Wesley.

Marcuse, Herbert. 1964. *One-Dimensional Man.* Boston: Beacon.

Mumford, Lewis. 1967–1970. *The Myth of the Machine.* Vol. 1: *Technics and Human Development.* Vol. 2: *The Pentagon of Power.* New York: Harcourt Brace Jovanovich.

Rostow, W.W. 1960. *The Stages of Economic Growth: A Non-Communist Manifesto.* Cambridge: Cambridge University Press.

Thurow, Lester C. 1996. *The Future of Capitalism: How Today's Economic Forces Shape Tomorrow's World.* New York: William Morrow.

Winner, Langdon. 1977. *Autonomous Technology: Technics-out-of-control as a Theme in Political Thought.* Cambridge, MA: MIT Press.

———. 1986. *The Whale and the Reactor: A Search for Limits in an Age of High Technology.* Chicago: University of Chicago Press.

————. 1993. "Social Constructivism: Opening the Black Box and Finding It Empty." *Science as Culture*, vol. 3, no. 16 (fall), pp. 427–52.

————. 1997. "Technological Determinism: Alive and Kicking?" *Bulletin of Science, Technology, and Society*, vol. 17, no. 1, pp. 1–2.

————. 1998. "Technology Today: Utopia or Dystopia?" *Social Research*, vol. 64, no. 3 (fall), pp. 989–1017.

2

Understanding Technological Culture through a Constructivist View of Science, Technology, and Society

WIEBE E. BIJKER

The term "social constructivism" has come to characterize much current thinking within the STS field. One of the chief proponents of this view has been Wiebe Bijker, professor of technology and society at the University of Maastricht in the Netherlands. Bijker first developed his constructivist line of thinking in a historical-sociological assessment of the development of the bicycle, which was published as part of a collection of essays he co-edited with Trevor Pinch and Thomas Hughes, *The Social Construction of Technological Systems* (1987), a book that has since become a benchmark in the field. He is also the author of *Of Bicycles, Bakelites, and Bulbs: Toward a Theory of Sociotechnical Change* (1995).

In this chapter Bijker argues that because "we live in a technological culture," we have an obligation to try to *understand* [that] technological culture." At the same time he wants to *politicize* it—that is, "to make explicit the political dimensions of the role of science and technology, to question the self-evident character of technological culture, and to put science and technology on the public agenda for political deliberation." Finally, he hopes to *democratize* modern scientific and technological culture by engaging more citizens in such political deliberation.

Not only is the centrality of Bijker's view important to

understanding STS as a field of study, but it is this very same constructivist view that helps to justify his and other activist arguments that citizens have both a democratic right and a responsibility to participate in the sociopolitical decision-making process so central to contemporary technological culture. In reading Bijker's essay, we should consider whether the constructivist view offers a clearer and more helpful understanding of science and technology than the view of science as objective and value free, or of technology as autonomous. We should also attempt to assess whether such a constructivist framework truly offers, as Bijker suggests, an opportunity for enhanced democratic participation in scientific-technical decision-making.

We Live in a Technological Culture

We live in a technological culture—in a culture that is thoroughly influenced by modern society and technology. It is thus not easily possible, I will argue, to understand modern Western culture without taking into account the role of science and technology. Indeed, this pertains to all aspects of this culture, not only to those that are openly linked to technology and science, such as communication, mobility, and environmental problems. Also other aspects of our culture are infused with science and technology—for example *language* (think of the common usage of metaphors derived from communication and computer technology); and *norms and values* (think of the differentiation of norms as to whether someone is "really" dead, as a result of the increased sophistication of organ transplant technologies); and *identity* (think of all the technological ways in which one's identity is defined: credit cards, health registrations, type of motorcar). It is to summarize this observation regarding the pervasiveness of science and technology in modern Western culture, that I use the sloganlike phrase "we live in a technological culture." It is not to argue that technology and science are the only important, or even the most important, aspects of our culture, but it is to argue that we cannot hope to understand modern culture without taking into account science and technology. It is, in other words, arguing for the pertinence of science-technology-society (STS) studies.

In this chapter I will introduce the constructivist perspective in STS studies. Besides briefly outlining the perspective itself, I will also address the wider issue of the future intellectual and political STS agenda, and the pertinence of a constructivist approach for such an agenda.

Understanding, Politicizing, and Democratizing Technological Culture

I want to argue that all who live in this culture—citizens, scientists, engineers, STS-students—have an obligation to try to *understand* the technological culture. This is basically the nineteenth-century enlightenment ideal. And, as argued above, we do need insight into the interplay of science and technology in society for such an understanding. My further goal, but not one which all will share, is to *politicize* technological culture: to make explicit the political dimensions of the role of science and technology, to question the self-evident character of technological culture, and to put science and technology on the public agenda for political deliberation. A possible third goal is to *democratize* technological culture: to promote particular normative choices of democracy when engaging in the debate on politicization. Even though there will be least agreement as to this last goal—people are likely to disagree about the particulars of their democratic ideals, as well as about the strategies to realize them—I will return to this issue in the conclusion. The main body of the essay is meant to provide a framework within which to understand technological culture, one with a specific sensitivity to its political aspects.

The Standard View of Science, Technology, and Society

Before presenting the constructivist framework, it is helpful to briefly discuss its counterpart, the standard image of science and technology—an image still widely held by citizens, students, and practitioners. In the standard image of science, scientific knowledge is objective, value-free, and discovered by specialists. Technology, similarly, is a rather autonomous force in society, and technology's working is an intrinsic property of the technical machines and processes. The left column of Table 2.1 summarizes this.

Some of the implications of these standard images are positive and comforting. Thus, for example, scientific knowledge appears as

Wiebe E. Bijker

a prominent candidate for solving all kinds of problems. In the domain of political thought, this naturally leads to "technocracy"-like proposals, where technology is viewed as a sufficient end in itself and where the values of efficiency, power, and rationality are valued independent of context. The standard view accepts that

Table 2.1
Standard and Constructivist Images of Science and Technology

Standard *View of Science and Technology (and Society)*	Constructivist *View of Science and Technology (and Society)*
Clear distinctions between the political and the scientific/technical domain	Both domains are intertwined; what is defined as a technical or as a political problem will depend on the particular context
Difference between "real science" and "trans-science"	All science is value-laden and may—again depending on the context—have implications for regulation and policy; thus there is no fundamental difference between "real science" and "trans-science," "mandated science," or "policy-relevant science"
Scientific knowledge is discovered by asking methodologically sound questions, which are answered unambiguously by nature	The stabilization of scientific knowledge is a social process
Social responsibility of scientists and technologists is a key issue	Development of science and technology is a social process rather than a chain of individual decisions; political and ethical issues related to science therefore cannot be reduced to the question of social responsibility of scientists and technologists
Technology develops linearly, e.g., conception → decision → operation	Technology development cannot be conceptualized as a process with separate stages, let alone a linear one

(continued on next page)

technology can be applied negatively, but in this view the users are to be blamed, not the technology. Not surprisingly, the standard image also leaves us with some problems. For some questions, for example, we do not yet have the right scientific knowledge. Also an adequate application of knowledge is, in this view, a separate problem. The role of experts is problematic in a specific way: How can experts be recognized by nonexperts?; How can nonexperts trust

Table 2.1 *(continued)*

Standard *View of Science and Technology (and Society)*	Constructivist *View of Science and Technology (and Society)*
Distinction between technology's development and its effects	The social construction of technology is a process that also continues into what is commonly called its "diffusion stage"; the (social, economic, ecological, cultural . . .) effects of technology are thus part of the construction process and typically have direct vice versa implications for technology's shaping
Clear distinction between technology development and control	Technology does not have the context-independent status that is necessary to hope for a separation of its development and control; its social construction and the (political, democratic) control are part of the same process
Clear distinction between technology stimulation and regulation	Stimulation and regulation may be distinguishable goals, but need not necessarily be implemented separately
Technology determines society, not the other way around	Social shaping of technology and technical building of society are two sides of the same coin
Social needs as well as social and environmental costs can be established unambiguously	Needs and costs of various kinds are also socially constructed—depending on the context, they are different for different relevant social groups, varying with perspective

the mechanisms that are supposed to safeguard the quality of the experts?; And, finally, how can experts communicate that esoteric knowledge to nonexperts? In the realm of technology, an additional problem is that new technologies may create new problems (which, it is hoped, in due time will be solved by still newer technologies).

Acceptable solutions for solving these problems are well known, up to the point of being trivial: more scientific and technological research, peer review, scientific expert advisory committees, and technology assessment. It is equally clear, however, that these "solutions" do not lead to as complete a disappearance of problems as the standard image of technology suggests. In the next section I will present an alternative image of science and technology, one which will yield some implications for understanding and politicizing technological culture.

A Constructivist View of Science, Technology, and Society

In the 1970s and 1980s detailed empirical research on the practices of scientists and engineers led to the formulation of a constructivist perspective on science and technology. This work by sociologists, historians, and philosophers became known under the banners of "sociology of scientific knowledge" (SSK) and "social construction of technology" (SCOT). I will briefly introduce both.

The Sociology of Scientific Knowledge (SSK)

Scientific facts are not found, literally dis-covered, in nature, but they are actively construed by scientists.[1] Readings from instruments do not speak for themselves but need to be constructed into scientific facts by researchers. The processes in which this is accomplished are social by their very nature: human researchers interacting with each other. They cannot be understood as mechanically following methodological rules; if that were so, we could replace scientists by computers.

The key idea is that nature does not dictate scientific facts. The image of scientific research—that doing an experiment is asking a question upon which nature unambiguously shouts yes or no—is false. SSK-researchers can show the *interpretive flexibility* of observations and propositions: that other readings are possible. Which of the possible readings subsequently stabilizes into generally accepted knowledge is subject to social processes. That is not to say that scientific knowledge is irrational, or disorderly, or unrelated to scientific experiments. It is to say that in order to understand the outcome of scientific research, and especially scientific controversies, we should aim at finding regularities of a sociological nature.

SSK-research has produced a variety of such insights. The *experimenters' regress* is one such example (Collins, 1985). Think of an experiment to investigate gravitational radiation—a kind of radiation that is similar to light but produced by moving massive bodies rather than by moving electrons. (The existence of gravitational radiation is predicted by Einstein's general theory of relativity.) Suppose a controversy develops over the outcome of this experiment, for example, that gravitational waves do indeed exist and have a particular character. How do we resolve that controversy? By doing another experiment to test the first experiment! But then a controversy over that second experiment may develop, and so on ad infinitum. Collins coined this circular trap the "experimenters' regress." Experimental work can only be used to test something if a way is found to break out of this circle, for example, by having consensus about the existence of gravitational waves of a particular character on theoretical grounds.

The *splitting-and-inversion model* is another example of SSK insight (Latour and Woolgar, 1979). On the basis of their anthropological study in a California biochemistry laboratory, Latour and Woolgar conclude that the process of scientific discovery is one of splitting-and-inversion. During the process of "science in the making,"[2] there is no distinction between an object and the statement *about* that object—there merely is the statement. But at the moment of social closure, when scientific consensus is reached, *splitting* between the object and the statement occurs, and the scientific fact becomes a statement *about* some part of nature. Also at that moment *inversion* occurs: the arrows of time and causality are inverted and the object is seen as being previous to and, indeed, the source of the statement.

A third insight relates to the *political dimensions of scientific controversies*. If, for example, scientists argue about the safety of nuclear reactors, the standard image of science can only suggest that one of the conflicting parties is wrong and the others are the good guys—for scientific knowledge is, in this view, unambiguously dictated by nature, so what is there to argue about? In a constructivist view, controversy among scientists is quite normal. Science cannot deliver complete certainty. The standard view that science can deliver certainty entails, what Collins and Pinch call, "flip-flop thinking"—it is all good or all bad. They conclude:

> The trouble is that both states of the flip-flop are to be feared. The overweening claims to authority of many scientists and technologists are offensive and unjustified but the likely reaction, born of failed promises, might precipitate a still worse anti-scientific movement. Scientists should promise less; they might then be better able to keep their promises. Let us admire them as craftspersons: the foremost experts in the ways of the natural world (Collins and Pinch, 1993, p. 142).

The Social Construction of Technology (SCOT)

Since the 1980s, and building on the SSK work discussed above, sociological and historical stories have developed a constructivist analysis of technology in contrast to the standard image of technology that was largely "technological determinist." Social shaping models stress that technology does not follow its own momentum nor a rational goal-directed problem-solving path but is instead shaped by social factors. (See Table 2.1 for a summary of standard and constructivist images of science and technology.)

In the *social construction of technology approach* (SCOT),[3] relevant social groups are the starting point. Technical artifacts are described through the eyes of the members of these groups. The interactions within and among relevant social groups can give different meanings to the same. Thus, for example, a nuclear reactor may exemplify to a group of union leaders an almost perfectly safe working environment with very little chance of on-the-job accidents compared to urban building sites or harbors. To a group of international relations analysts, the reactor may, however, represent a threat through enhancing the possibilities of nuclear proliferation, while for the neighboring village the chances for radioactive emissions and the (indirect) employment effects may strive for prominence. As a workplace, nuclear technology is succeeding quite well; whereas, as a source for international tension or as an environmental hazard, it may be evaluated quite differently. This demonstration of *interpretive flexibility* is a crucial step in arguing for the feasibility of any sociology of technology—it shows that neither an artifact's identity, nor its technical "success" or "failure," are intrinsic properties of the artifact but subject to social variables.

The next step is to describe how artifacts are, indeed, socially constructed, thus tracing the increasing (or sometimes decreasing) degrees of stability of that artifact. The concept of "technological frame" is proposed to explain the development of heterogeneous socio-technical ensembles, thus avoiding social reductionism.

A technological frame structures the interactions between the "actors" of a relevant social group. A key characteristic of the concept is that it is applicable to all within the relevant groups—technicians and others alike.[4] It is built up when interaction "around" a technology starts and continues. Existing practice does guide future practice, though not completely deterministically. The concept of "technological frame" forms a hinge in the analysis of socio-technical ensembles: it sets the way in which technology influences interaction and thus shapes specific cultures, but it also explains how a new technology is constructed by a combination of enabling and constraining interactions within relevant social groups in a specific way.

The Politicization of Technological Culture

The constructivist conception of technology is, I want to argue, crucial for a discussion of the politicization of technology. The argument involves two steps. First, I will argue that a constructivist analysis, in some form, is a condition sine qua non for any politics of technology. This results in stressing the malleability of technology, the possibility for choice, the basic insight that things could have been otherwise. But technology is not only malleable and changeable—it can be obdurate, hard, and very fixed too. The second step, then, would be to analyze this obduracy of sociotechnical ensembles.

The constructivist perspective provides a rationale for a politics of technology. It does so by exemplifying the very possibility of a social analysis of technology. Demonstrating the interpretive flexibility of an artifact makes clear that the stabilization of an artifact is a social process, and hence subject to choices, interests, and value judgments—in short, to politics. Without recognizing the interpretive flexibility of technology, one is bound to accept a technologically determinist view. A technological determinist view does not stimulate citizens' participation in processes of democratic control of technology, since it conveys an image of autonomy and the impossibility of intervention.

Apart from having a role in the public debate about sociotechnical choices, to demonstrate the interpretive flexibility of sociotechnical ensembles is also crucial in a more analytical sense. For without such a perspective an analysis of technology and society is bound to reproduce the stabilized meanings of technical artifacts and will miss many opportunities for intervention. The interpretive flexibility of technology often will not be obvious and needs to be demonstrated in a rigorous way to escape the rather trivial level of observation that technology is humanmade, and hence subject to many societal influences. The constructivist argument is that the core of technology—that which constitutes its working—is socially constructed. This is a way to take up the challenge of Langdon Winner's observation that "artifacts have politics." Such a perspective seems necessary to overcome the standard view of technology and society, in which "blaming the hardware appears even more foolish than blaming the victims when it comes to judging conditions of public life" (Winner, 1986:20).

The Hardness of Facts and Machines

Let me now turn to the second step in the argument. To argue for the malleability of technology does not imply that we forget the solidity and momentum of sociotechnical ensembles. Such negligence might result in an equally counterproductive cultural-political climate, because it invokes too optimistic an expectation which in turn may cause disillusions. A politics and a theory of sociotechnology have to meet similar requirements here—a balance between actor and structural perspectives in the second. Sociotechnical ensembles not only have interpretive flexibility, they can also be fixed and obdurate, and they will accordingly function in the societal power struggles over technology.

We can distinguish two aspects of power—a *micropolitics of power*, in which technologies may be used as instruments to build up networks of influence, and a *semiotic power structure*, which results from these micropolitics and constrains actors (Bijker, 1995, chapters 4–5). The semiotic power originates from the fixity of meanings, which is built up during the formation of a technological frame as a result of the micropolitics of relevant social groups. The groups have, in building up the technological frame, invested so much into the key technology that this technology's meaning becomes fixed—it cannot be changed easily, and it forms part of an enduring network of practices, theories, and social institutions.

From this time on, it may indeed happen that, naively speaking, the technology "determines" social development. Such an "exemplary" sociotechnical ensemble is, at the same time, the result of micropolitical interaction processes and one of the elements of a semiotic power structure. A sociotechnical ensemble can also be an important boundary-creating instrument. Then it functions on the border between two relevant social groups, often especially in the hands of actors with a low inclusion in the respective technological frames.[5]

For the low included actors, such an artifact presents a "take it or leave it" choice—they have no chance of modifying the artifact when they "take" it, but life can go on quite well when they "leave" it. For the high included actors, on the contrary, there is no life without the exemplary artifact, but there is a lot of life within it. The obduracy of artifacts as boundary objects for low included actors consists in this "take it or leave it" character. For such actors, there is no flexibility; there is no differentiated insight; there is only technology, determining life to some extent and allowing at best an "all or nothing" choice. This is the obduracy of technology which most people know best. This is the kind of obduracy that gives rise to technological determinism. For high included actors, obduracy of technological ensembles presents itself as the technology being all-pervasive, beyond questioning, and dominating thoughts and interactions.

Artifacts as boundary objects result in obduracy because they link different relevant social groups together into a semiotic power structure. To make the "take it" choice with respect to such an artifact results in being included into such a semiotic power structure. This implies being subject to power relations that one would otherwise—in the case of a "leave it" choice—be immune to. Someone who buys a car, for example, is thereby included in the semiotic structure of automobiling: cars-roads-rules-jams-gasoline-prices-taxes. This will result in automobilists exerting power, for example by using the car during rush hour and thereby contributing to a traffic jam, but will also make them subject to the exertion of power by others—the traffic jam again. Without a car, however, jams and oil prices simply do not matter. "Exemplars," or key artifacts result in obduracy because they constitute to an important degree the world in which one is living. This also implies inclusion in a semiotic power structure but with different possibilities and effects. Many of the power interactions are now in terms of the exemplary artifact. Leaving the car standing is less likely an option, but changing one's driving hours or routes (to beat the jams), changing from gasoline to diesel or liquid gas (to beat the taxes), or changing to a smaller car (to reduce parking problems) are possibilities.

Different Kinds of Expertise

The issue of expertise lies at the heart of both the practice of doing STS and of a politicization of technological culture. Is the consequence of a constructivist view of scientific knowledge and technological devices—the view that the development of science and technology is a social process—that scientific expertise does not exist, or is irrelevant? This is not the case, neither for STS students, nor for participants in political debates about science and technology. I shall discuss both aspects in turn.

The expertise of any researcher, be it a physicist, a sociologist, or a philosopher, is formed and checked in a socialization process within the relevant scientific community. Key concepts to understanding this process are "peer review" and Kuhn's "paradigm" (1962). There is nothing different in the case of STS scholars. They too will acquire their expertise in undergraduate and postgraduate studies, being slowly socialized into the STS community. There is one difference, however, in contrast to other scientific disciplines. STS students study other sciences and engineering practices, which requires them to have a more than superficial knowledge of this other scientific or engineering discipline. As part of their research, STS researchers thus must also socialize a little into their object of study—that other community. This may seem rather self-evident, and no different from requiring an anthropologist who studies the culture of drug addicts to socialize into the drug scene, but in the field of STS this argument needs to be made explicitly. Until constructivist studies started in the 1970s, the view was generally held that science as an object of study was different from all other objects of study (such as the drug scene): the development of the *content* of science, of scientific knowledge per se, was not amenable to analysis by anyone other than the practicing scientists themselves. This is like arguing that the culture of the drug addicts can only be studied by the drug addicts themselves.

SSK changed this—the constructivists showed (as presented above) that the development of scientific knowledge is a social process and thus open to analysis by sociologists. The implication is, of course, that students need to familiarize themselves thoroughly with the culture that is the object of their study. In other words, the STS student must be prepared to acquire detailed scientific and technical knowledge, just as the anthropologist must be prepared to acquire such knowledge about the practices of drug dealing and use.

The reason for an STS student to acquire detailed scientific and technical knowledge that I just discussed is a methodological one—only on the basis of such knowledge can proper STS research thrive. There is as well, however, an additional, more political reason to stress this need. Without such detailed knowledge, STS scholars cannot claim any special authority to engage in discussions about the societal role of science and technology, and about the politics and policy of science and technology. This brings me to the second question of expertise—expertise in relation to the politicization of technological culture.

What about the expertise needed by other relevant social groups to engage in political debates about science and technology? Two arguments are relevant, and they define both extremes of a spectrum of answers to this question. The first argument is based on the standard image of science and technology. It runs as follows: science and technology are special domains, and the expertise of scientists and engineers is needed to discuss its development. There is a clear distinction, however, between the contents of science and technology and their applications. About these applications other social groups of citizens and politicians can debate.

The second argument is based on a naive and extreme form of social constructivism and runs as follows: science and technology are in no way different from other domains, and there is no reason to give the expertise of scientists and engineers any special status. I think this latter argument is as foolish as the first. The constructivist perspective of science and technology, the one that I have presented in this essay, does not support the view that scientific expertise is nonexistent, or irrelevant, or identical to the expertise of any nonscientist.

I want to conclude that a constructivist view of knowledge and technology implies the existence of *a variety of expertise*. Different relevant social groups have their specific kinds of expertise—we are all experts in specific ways. Note the "specificity" condition: scientists have their own invaluable form of expertise, as do STS scholars, and also groups of citizens, politicians, and other experts. I am not arguing that an average citizen is able to design a nuclear reactor or a river dike, but I am arguing that more is involved in designing large projects such as nuclear power stations and water management systems than is described in the engineers' handbooks. For those aspects, others are experts and need to be involved, and they need to be involved *in the whole design process* in as early a stage as possible.

Conclusion: Learning Programs
for Citizens, Scientists, and STS Students

Let me, by way of conclusion, briefly turn to the issue of democratization of technological culture. First of all, it is important to recognize that the concept of democracy can have different meanings in different cultures—for example, the Jeffersonian ideal of direct democracy is quite different from the representational elitist European ideal of democracy, a distinction that will yield very different proposals for a democratization of technological culture.[6] Transcending these differences, I think that it is possible to sketch some of the consequences of the view that I have presented in this essay as to the way in which various groups should relate to the issue of democratizing technological culture.

For citizens the consequence is that more educational time should be spent on scientific and technical literacy, in combination with STS teaching. The first, without the latter, is only useful for future scientists, not for future citizens. It also means that citizens can and should be informed about and involved in discussions about scientific and technological developments, although the precise form will depend on the model of democracy that one supports. In such a technological culture, scientists and engineers are expected to engage with other social groups about their work. This implies that also they should be taught STS insights to enable them to reflect upon their work and its implications for society. STS students themselves should continue to engage in science and engineering practices, partly as a prerequisite for their research and partly as a contribution to developing a politicization of technological culture.

Notes

1. See Collins and Pinch (1998b) and Collins and Pinch (1998a) for a good first-level introduction to SSK. See Collins and Latour (1987) for more sophisticated discussions of various constructivist views of science.

2. This phrase is coined in Latour (1987).

3. See Bijker (1995) for a full account of this social construction of technology approach.

4. It is on this point that "technological frame" differs crucially from concepts such as Kuhn's (1962) "paradigm" or Dosi's (1982) "technological paradigm."

5. An actor's inclusion in a technological frame determines the degree to which this frame guides the thinking and interacting of the actor. A high included actor works and thinks very much in terms of the technological frame; an actor with a low inclusion much less so.

6. See Wilde (1997) for a critique of the naivety of STS scholars in this respect.

References

Bijker, Wiebe E. 1995. *Of Bicycles, Bakelites and Bulbs: Toward a Theory of Sociotechnical Change, Inside Technology*. Cambridge, MA: MIT Press.

Bijker, Wiebe E., Hughes, Thomas P., and Pinch, Trevor, eds. 1987. *The Social Construction of Technological Systems: New Directions in the Sociology and History of Technology*. Cambridge, MA: MIT Press.

Collins, Harry. 1985. *Changing Order: Replication and Induction in Scientific Practice*. London: Sage.

Collins, Harry, and Pinch, Trevor. 1993. *The Golem: What Everyone Should Know About Science*. Cambridge: Cambridge University Press.

———. 1998a. *The Golem: What You Should Know About Science*. 2nd ed. Cambridge: Cambridge University Press.

———. 1998b. *The Golem at Large: What You Should Know About Technology*. Cambridge: Cambridge University Press.

Dosi, Giovanni. 1982. "Technological Paradigms and Technological Trajectories: A Suggested Interpretation of the Determinants and Directions of Technical Change." *Research Policy*, vol. 11, no. 3, pp. 147–62.

Kuhn, Thomas. 1962. *The Structure of Scientific Revolutions*. Chicago: University of Chicago Press.

Latour, Bruno. 1987. *Science in Action: How to Follow Scientists and Engineers Through Society*. Cambridge, MA: Harvard University Press.

Latour, Bruno, and Woolgar, Steve. 1979. *Laboratory Life: The Social Construction of Scientific Facts*. Beverly Hills, CA: Sage.

Wilde, Rein de. 1997. "Sublime Features: Reflections on the Modern Faith in the Compatibility of Community, Democracy, and Technology." Pp. 29–49 in Sissel Myklebust, ed., *Technology and Democracy: Obstacles to Democratization—Productivism and Technocracy*. Oslo: TMV, University of Oslo.

Winner, Langdon. 1986. *The Whale and the Reactor: A Search for Limits in an Age of High Technology*. Chicago: University of Chicago Press.

3

Three Perspectives in STS in the Policy Context

LARS FUGLSANG

Lars Fuglsang teaches science and technology policy studies at Roskilde University in Denmark, a country that has developed a strong tradition of public participation in technology assessment. He is the author of *Technology and New Institutions: A Comparison of Strategic Choices and Technology Studies in the United States, Denmark, and Sweden* (1993), which argues that technology studies is a mosaic of approaches, rather than one singular method.

In the present essay, Fuglsang defends what he terms the "interactive" view of the STS relationship. In this view—to which the research of Bijker and other social constructivist scholars has contributed—science, technology, and society are seen as interrelated parts of a "seamless web," rather than as separate entities accidentally joined. He contrasts this more holistic or interactionist view with the depiction of science and technology as deterministic forces of socioeconomic change, and the idea that social forces are the shapers of science and technology. For Fuglsang, the arrow of determination is not one-way in either direction. While the notion of two interacting arrows may seem incongruous and in apparent conflict, Fuglsang argues that, in fact, they are really more representative of different phases of techno-scientific development in which science and technology can display both flexibility and inflexibility.

Thus, for Fuglsang, it is important to recognize how different STS perspectives may be appropriate to different stages of
technological development. It is in the earliest stages of development that most technologies retain the most flexibility, so that the
arrow of influence moves from society to technology. In later
stages technologies become more entrenched, and the arrow of
influence can move back the other way. In reading Fuglsang's
essay, we may thus ask ourselves whether Fuglsang provides a
way to harmonize the arguments for technological determinism
found in Winner with the more constructivist approach to technology studies found in Bijker.

We might also consider, however, whether such a conceptual
harmonization is sufficient to the challenge of the social control of
technology. David Collingridge (1980), for instance, argued that
there is a fundamental paradox for any attempt to control technology; that is, early in the development of any technology, when
it is relatively easy to exercise some influence over it, we seldom
know enough to actually do so, but once the technology has been
socially deployed and we know enough about its negative consequences to want to control it, it has become rather difficult to do
so. Does Fuglsang provide any way to deal with such a dilemma?

In this essay I want to distinguish among what I view as three
main perspectives in STS regarding science and technology and
their relation to society. These three perspectives are: 1) science
and technology shape society; 2) society shapes science and technology; and 3) an interactive view of the science, technology, and
society relationship. Each of these perspectives developed over the
course of a ten- to twenty-year period after World War II, and each
is tied to a specific policy context. However, they can also be seen as
simultaneous perspectives that have been competing with one
another during the past fifty years. For present purposes, however,
I treat them historically in the policy context.

Science and Technology Shape Society

This perspective evolved out of an optimistic atmosphere of
science and technology in the years after World War II. Science and
technology were, for the first time in modern history, considered as
forces of socioeconomic change that made a difference for society
and the economy.

One important contribution to this perspective was made by Vannevar Bush, the U.S. president's advisor for scientific research and development. His 1945 report, "Science, the Endless Frontier," is a source of inspiration for the modern funding system for science. Bush argued for a "basic science" which would eventually have positive consequences for the economy. In his vision, science should not be targeted directly by the government; rather, funding should be determined by scientists themselves through a system of peer review.

In this, Bush agreed with other authors of the time. During the 1930s, the British scientist John Desmond Bernal had already argued for science as a cornerstone in the building of modern society (1939). He also believed that science should be protected from direct external interference by society. Such considerations were also important for Derek de Solla Price (1963), one of the founding fathers of science studies. De Solla Price wanted to create a "science of science" that could lead to improvements of science institutions.

Arguments about the important role of science and technology were also, directly or indirectly, given by the Nobel Prize Laureates in economics Kenneth Arrow (1962) and Robert M. Solow (1956). Arrow argued that science needs public support because it is associated with economic risk (market failure). Thus, according to Arrow, science is characterized by its fundamental "uncertainty" (its results cannot be predicted), its "indivisibility" among economic actors (it cannot be divided among them like a cake), and its difficulty to be "appropriated" economically (since knowledge easily flows from one actor to another—the imitation problem). The existence of high economic risk, he argued, implies underinvestment in science by private firms, thus providing a rationale for public support.

Solow was investigating growth in output per worker in the U.S. in the period from 1909 to 1949. He found that 87.5 percent of the growth could not be explained by an increase in capital per worker as was usual, but had to be assigned a residual factor he called "technology." This was not very convenient, since economic theory had not generally treated technology in these models. Nonetheless, the argument suited the spirit of the time well— that science and technology were forces contributing to growth and prosperity.

Another variant of the "science and technology shape society" approach emerged from economic-historical theories. This variant has its origin in the work of the Russian economist N.D. Kondratiev (1935) who found statistical evidence for long-term economic cycles of forty-five to sixty years in the period from the eighteenth century through to the 1920s in the U.S., U.K., and France. Building on Kondratiev's work, the Austrian economist Joseph Schumpeter attempted to explain the upswings and downswings in the economy with entrepreneurship and innovation. Entrepreneurs played, according to Schumpeter (1939), a creative-destructive role for economic recovery. Extending these arguments, Christopher Freeman and Carlotta Perez (1988) have put forward the idea that economic and institutional development is motivated by shifts in techno-economic paradigms, each containing a new key technology, such as the steam engine (late eighteenth century), railways (mid-nineteenth century), electricity (late nineteenth century), petrochemicals (early twentieth century), and information technology (mid-twentieth century).

This is a fairly broad but attractive approach that nicely chains together economic analysis with sociology and history. Similar ideas have been developed by Giovanni Dosi (1982), who argues that technology within each techno-economic paradigm evolves along certain trajectories (defined as "the pattern of 'normal' problem-solving activity [i.e., of progress]"). One example is the trajectory of semiconductors leading toward smaller, cheaper, more reliable, higher memory computer chips.

The "science and technology shapes society" approach also comprises more pessimistic views, such as those proposed by the French sociologist/philosopher Jacques Ellul (1964) or the German philosopher Jürgen Habermas (1973). Here, in an emerging tradition of "philosophy of technology," focus is on the alienating effects of science and technology on human life or other philosophical aspects of science and technology. Science and technology inform an instrumental rationality or a technological regime that overshadows and represses other equally important aspects of human life, such as philosophical or religious thought. The approach can be seen as a sophisticated historical-philosophical critique of civilization—one which is also elaborated by authors more directly linked with STS, such as Lewis Mumford (1967, 1970) and Langdon Winner (1986).

The "science and technology shapes society" perspective is sometimes labeled a "technological determinist approach." This can be misleading, however, since technology is seldom seen as an autonomous force of its own right, but more as a normative choice of Western society in a broad sense (see Habermas, 1973). Where and how one should apply the term *technological determinism* is discussed by Bruce Bimber (1994).

Society Shapes Science and Technology

The "society shapes science and technology" perspective turns things around. Now, the determinant force is not technology but society. This approach has its origins in pressures from both business and academic discussions during the early 1970s. In business, there was a growing concern that science should be more directly connected to commercial purposes. The spill-over from basic research to business in general was perceived as poor. "Contracted research" instead of "basic research" was suggested (see the British report by Lord Rothschild on "The Organization and Management of Government R&D" in Seal, 1971; see also Elzinga, 1988).

The academic discussions emphasized social forces external to science and technology. Science and technology were to be seen in the light of the social, economic, and political interests and concerns of the wider population. There were two special lines of reasoning. One argued in strategic and political terms that science and technology should be more explicitly linked to social forces, while the other emphasized a sociological and academic approach, seeking to examine and conceptualize links between social forces and science/technology.

Thus, the OECD "Brooks Report" of 1971 (Brooks, 1971) pleaded for incorporation of "strategic choices" into science and technology—i.e., integration of the social concerns of civil groups. Furthermore, in 1972, the Office of Technology Assessment was established under the auspices of the U.S. Congress. Its main function was to identify the impact of technological application and thereby to support political deliberations concerning science and technology.

The sociological/academic line was pursued, among others, in the so-called strong program of "the sociology of scientific knowledge" (SSK) in the U.K. The idea was to study moments of alternative opportunity in science and thereby show how competing options were linked to priorities of different social forces. This type of argument was behind several valuable contributions—such as David Noble's *Forces of Production* (1984)—analyzing the development of the numerically controlled machine tool, and the publication edited by Donald MacKenzie and Judy Wajcman, *The Social Shaping of Technology* (1985).

In some Scandinavian variants of the perspective, such as the DEMOS and the UTOPIA projects, there were attempts to convert these insights into "action research." Relevant social forces were to be mobilized and empowered, especially in newspaper typesetting, to be able to influence technical change according to their interest. The Scandinavian approach was inspired by industrial sociology. For example, Harry Braverman (1974) argued that new technology was often applied by industrialists to control work, which also lead to labor deskilling. But the Scandinavian action research projects were far more pragmatic and constructive than Braverman's analysis would suggest. DEMOS and particularly UTOPIA sought to preserve and make use of workers' skills, and to improve the quality of work during technical change in a way that was operational on both sides of the table.

The main idea of the "society shapes science and technology" approach was thus to see technology as open to external forces and negotiation. Analysts stressed that technical change is not neutral but biased by social and economic forces. Civil groups could and should be empowered and integrated into decisions concerning science and technology. In business this perspective was reflected in greater pressure for more relevance in science.

The Interactive View of the Science, Technology, and Society Relationship

The interactive perspective was presented by, among others, Wiebe Bijker (1987), Bruno Latour (1987), and Michel Callon (1986). The initial steps were taken during the 1980s, but the breakthrough for the approach came in the early 1990s under such headings as "the social construction of technology" (SCOT) and "actor-network-theory." SCOT is closely associated with SSK but shifts the focus from science to technology.

Here, technology is seen as having "interpretive flexibility," which implies that it does not develop in a linear way. Rather, technical change contains, like science in the SSK program, moments of alternative possibilities. Which steps are taken in technology depends on the specific social constituencies that are involved with the technology.

The most cited example of this approach is "the social construction of the bicycle" by Trevor Pinch and Wiebe Bijker (1984; further developed in Pinch and Bijker, 1987). The analysis goes like this: In the late nineteenth century, three competing bicycles were conceived, one made for macho men (a risky bike to ride, with a high front wheel and a low back wheel), one for women (with pedals on the same side of the bike, for example, to solve the dressing problem and meet moral standards), and a practical bicycle, mostly for elderly men. Each bicycle was, in the beginning, equally important. (All this occurred despite the fact that Leonardo da Vinci had already designed the bicycle as we know it today in the fifteenth century.) It was only when the rubber tire and improved brakes were created that the modern bicycle started to catch on. The reason was not necessarily that it was better, but, according to Pinch and Bijker, that a compromise could now be produced between the groups of elderly men (in need of a practical bike) and macho men (who could now have a challenging "fast" bike instead of a risky bike).

From this case study, Pinch and Bijker try to ground a theory comprising several concepts such as "interpretive flexibility" and "the relevant social group" (elderly men, macho men, and the like). The relevant social group is probably meant to be a more specific category than the hitherto dominant broad notions of society. They also make the argument that the relation between society and technology is not one between two distinct entities (society versus technology) but is rather a "seamless web." In addition to these concepts, Bijker has tried to conceptualize institutional and other constraints on technologists by the term "technological frame" (see Bijker et al., 1987). How technological frame is related to concepts of institutions more generally, or to notions of social structure, is not so clear. Bijker, however, in his latest work (Bijker, 1995), has tried to go some steps further along these lines by examining the concept of power.

The Bijker approach has had far-reaching consequences for STS. Together with other similar approaches, such as the actor-network theory of Latour and Callon, it has informed a refreshing methodological discussion in STS. Criticisms have been raised from the "science and technology shape society" perspective, most sharply by Langdon Winner (1993). Winner sees SCOT as voluntaristic, with a naive and relativist conception of reality (like "anything goes"). Another problem of the SCOT approach is its tendency to build up theory from single cases. Other relevant social science theory and method is hardly scrutinized or applied, and statistical analysis is absent. As a consequence, the concepts developed may not seem very solid. SCOT is an intelligent and refreshing approach that has significantly improved the intellectual capability of STS, but it is not necessarily a basis for further investigations. Its long-term influence may be more indirect.

Policy, Power, and Method:
The Three Perspectives Compared

A summary comparison of the three perspectives are presented in Table 3.1. As indicated, the three perspectives on science, technology, and society can be connected to competing views of policy, power, and method. In the "science and technology shape society" perspective, policy is "for" or "against" science and technology. Policy can serve to protect science and technology from external interfer-

Table 3.1
Three Perspectives on STS Compared

	Science and technology shape society	Society shapes science and technology	Interactive approaches
Time	1950s–60s	1970s–80s	1990s
Definition of technology	Cause	Consequence	Cause and consequence
Independent variable	Technology	Society	Social group
Relation of actor to technology	Beneficiaries (or victims)	Negotiate interests	Seamless web
Role of policy	Protect or reject science and technology	Empower actors, create networks	Democratize
Power structure	Technological regime	Negotiations	Frames, discourses
Method	Study impact technology	Follow the artifact	Follow the actor

ence, and seek to improve institutions of science and technology.

Alternatively, policy is seen in the Luddite tradition as coming from below, from critical social groups that want to slow down the pace of technical development (the Luddites were a British workers' movement that destroyed textile machinery from 1811 to 1817 in order to slow down technical change). Power is inherent to the "technical regime," through peer-reviewed funding systems, for example. The preferred method of study is to examine the impact of science and technology on society, either the economic impact (its correlation with economic growth) or the social impact (its social context).

In the "society shapes science and technology" perspective, policy is understood as networking and strategic interaction among concerned social groups. From a practical point of view, the goal is to integrate actors, to empower them to formulate views on science and technology, and to involve them in the implementation process as well. This is intended not only from a critical point of view, in support of employees, for example, but also in policy. Most European technology policy programs in the 1980s, such as the ESPRIT program of the European Union, were established through active involvement of concerned actors in the policy process. The twelve largest electronics firms in Europe created the ESPRIT program, and these firms also received the major part of the ESPRIT funding in its early years. Hence, power is negotiated among concerned social and economic groups. The appropriate method is to follow the artifact, and from that perspective to identify relevant actors.

In the interactive perspective, one major concern is the democratization of science and technology. Science and technology are seen as social relations and thus open to discussions of all kinds. Power has to do with the social and civil discourses that surround technology in a much more radical and fundamental sense than in the strategic approach. However, any attempt to democratize these discursive processes is confronted by major practical problems. It requires the building of sophisticated and complicated new institutions that are legitimate as democratic institutions. These are problems that SCOT shares with other recent mixed approaches to democracy in political science, such as the *associative democracy* approach (Hirst, 1993), for example, putting confidence in new forms of associations among public organizations, private firms, and civil society. The interactive approach is much more precise when it comes to its primary methodological suggestion, to "follow the actor."

Phases of Technology

Superficially, at least, the three STS perspectives appear to be in conflict, as they present incongruous views on the subject matter, methodology, and the purpose and relevance of theory. The first perspective has been accused of being determinist or technocratic, the second of being radical or business-targeted, and the third of being voluntaristic or relativist. Upon closer examination, however, the three different approaches are not really incongruous at all. They simply deal with different aspects of science and technology, and may be combined if we divide specific development or technical change into different phases. By phases I mean the different stages of innovation of a technology (or piece of science), somewhat similar to the concept of the product life cycle—though with emphasis on technological rather than commercial aspects of development. For the purpose of this essay, it is useful to distinguish between three such phases:

The phase of flexibility: This phase refers primarily to the initial stages of technological innovation. Here, the final form of the technology is not yet established. As in the cases of, for example, the creation of the railroad system, radio, television, the cassette tape, the video tape, and the computer, there were a number of competing technical solutions in circulation. The initial flexible stage of technical change is normally characterized by many failing experiments and moments of alternative possibilities. Public funding may, in some cases, be needed to bear the development costs, although large companies of the capital goods sector will often be the key players and will conduct the critical experiments and cover the costs.

The phase of momentum: This term refers to the phase when a particular technology has gained strength and widespread acceptance, while others have been excluded (see, e.g., Hughes, 1969 and Staudenmaier, 1986). Because modern technical systems are complex, and incorporated into the routines and practices of many employees and firms, they will, at this stage, be linked to "vested interests" as they start to gain momentum. Once the crucial decisions have been taken, the technology cannot be changed easily or without major costs. A trajectory of "normal" problem-solving is starting to take form, along which the technology is now further developed.

The phase of diffusion: In this phase, the artifact has matured and is diffused to consumer industries and applied by final users. At this stage, a "reversed product life cycle" may, however, take off in some industries, particularly in our present age of information technology. A reversed product life cycle will move from changes of process to changes of product rather than the other way around. Thus, according to a theory proposed by Richard Barras (1986), information technology applied in services first leads to rationalization of labor (cost-saving activity), then it eventually enables development of qualitatively new production systems (using the information technology for new purposes), and finally, as a result of this, to the innovation of new products (when the final user starts to see new product characteristics). The cycle is reversed in another sense also, since some of the new products may create higher flexibility in production and consumption. Hence, a new round of flexibility evolves—valid for some employees and customers at least—leaving space for further empirical investigation.

From this analysis we may draw a number of lessons about science, technology, and society. First, science and technology are both flexible and inflexible, depending on the stage of development and the industry involved. Hence, the design of the computer was in the beginning "interpretively flexible." Eventually, however, it became more standardized, due to network externalities and economics of scale. Standardized information technology is, however, the basis for new products in services and manufacturing industries of which some are relatively flexible. Computers, understood as materializing social relations, are therefore both shaping and—still—shaped by society.

Second, science and technology have an ambiguous role in economic development. Science and technology initiatives play, as Schumpeter said of entrepreneurs, a creative/destructive role. In the case of the computer, for example, it causes unemployment and deindustrialization but also new socioeconomic opportunities. Further down the line, the technology enables new services and new flexible relationships between service workers and customers.

Third, technical change can be democratized at many stages, from the creation of a capital good to the final diffusion of it to the service sector. But the issue of democracy is very complicated because many powerful actors are at play. They often operate on a global scale and interact in networks, although not clearly within the boundaries of a national legal framework, however. Therefore, actors are not easily submitted to national legislation. Representative democracy, as we know it, seems to be a necessary—if unsatisfying—condition for technology and democracy.

The relationship of technology and democracy is a very complicated and important issue for the future. I offer here a brief list of some mechanisms that, in my opinion, are essential for technology and democracy at national levels and that could be examined historically:

1) Each country may create mechanisms through which users of technology can express their dissatisfaction in the early stages of technological development and diffusion. The Danish idea of consensus conferences may provide one example of this. The consensus conference is an organized discussion among lay people and experts under the auspices of the Danish parliament that leads to a consensus report. 2) Consumer movements can be stimulated to represent consumer interests at an aggregated level. 3) The population can be educated and empowered to take part in discussions of technical change at all possible levels.

Finally, services play a crucial role for applications of information and communication technology. This is an issue that until now has been underresearched. Because services often take the form of relations rather than products, the diffusion of technology to services and innovation in services becomes a focal point for studying the impact of technology on human interaction.

Concluding Remarks: Opening the Doors of STS

In this essay I have suggested that science, technology, and society (STS) consists of several seemingly competing, if not conflicting, perspectives, because they relate to different notions of power, policy, and method. Nevertheless, the perspectives can be combined. Combining the perspectives does not mean, however, that we create a unitary approach of STS. What I intend is rather a pluralistic and open approach. To open the doors among the different perspectives is a major challenge for STS, which may also require a thorough deliberation of the different related policy interests. It may be more comfortable to remain within one of the perspectives, but to move across their thresholds can lead to more fruitful scholarly interaction and a stronger role for STS, which, without such movement, may run the risk of being pulled apart by the competing policy interests.

References

Arrow, Kenneth. 1962. "Economic Welfare and the Allocation of Resources for Invention." Pp. 609–26 in *The Rate and Direction of Inventive Activity: Economic and Social Factors.* Princeton, NJ: Princeton University Press.

Barras, Richard. 1986. "Towards a Theory of Innovation in Services." *Research Policy*, vol. 15, no. 4 (August), pp. 161–73.

Bernal, John Desmond. 1939. *The Social Function of Science.* New York: Macmillan.

Bijker, Wiebe E. 1995. *Of Bicycles, Bakelites, and Bulbs: Toward a Theory of Sociotechnical Change.* Cambridge, MA: MIT Press.

Bijker, Wiebe E., Hughes, Thomas P., and Pinch, Trevor, eds. 1987. *The Social Construction of Technological Systems: New Directions in the Sociology and History of Technology.* Cambridge, MA: MIT Press.

Bimber, Bruce. 1994. "Three Faces of Technological Determinism." Pp. 79–100 in Merritt R. Smith and Leo Marx, eds., *Does Technology Drive History? The Dilemma of Technological Determinism.* Cambridge, MA: MIT Press.

Braverman, Harry. 1974. *Labor and Monopoly Capital: The Degradation of Work in the Twentieth Century.* New York: Monthly Review Press.

Brooks, Harvey. 1971. *Science, Growth and Society: A New Perspective.* Paris: Organization for Economic Cooperation and Development.

Bush, Vannevar. 1945. *Science, the Endless Frontier: A Report to the President on a Program for Postwar Scientific Research.* Washington, DC: Office of Scientific Research and Development, U.S. Government Printing Office.

Callon, Michel. 1986. "The Sociology of an Actor-Network: The Case of the Electric Vehicle." Pp. 19–34 in Michel Callon, John Law, and Arie Rip, eds., *Mapping the Dynamics of Science and Technology: Sociology of Science in the Real World.* Basingstoke, Hampshire: Macmillan.

Collingridge, David. 1980. *The Social Control of Technology.* New York: St. Martin's Press.

Dosi, Giovanni. 1982. "Technological Paradigms and Technological Trajectories: A Suggested Interpretation of the Determinants and Directions of Technical Change." *Research Policy*, vol. 11, no. 3, pp. 147–62.

Ellul, Jacques. 1964. *The Technological Society*. Trans. John Wilkinson. New York: Knopf (French original, 1954).

Elzinga, Aant. 1988. "From Criticism to Evaluation." Pp. 29–58 in Andrew Jamison, ed., *Keeping Science Straight*. Gothenburg: Department of Theory of Science, University of Gothenburg.

Freeman, Christopher, and Perez, Carlotta. 1988. "Structural Crises of Adjustment: Business Cycles and Investment Behaviour." Pp. 38–66 in Giovanni Dosi, Christopher Freeman, Richard Nelson, Gerald Silverberg, and Luc Soete, eds., *Technical Change and Economic Theory*. London: Printer Publishers.

Fuglsang, Lars. 1993. *Technology and New Institutions: A Comparison of Strategic Choices and Technology Studies in the United States, Denmark, and Sweden*. Copenhagen: Academic Press.

Habermas, Jürgen. 1973. *Erkenntnis und Interesse*. Frankfurt: Suhrkamp.

Hirst, Paul. 1993. *Associative Democracy: New Forms of Economic and Social Governance*. Oxford: Polity Press.

Hughes, Thomas P. 1969. "Technological Momentum in History: Hydrogeneration in Germany 1898–1933." *Past and Present*, no. 44 (August), pp. 106–32.

Kondratiev, N.D. 1935. "The Long Waves in Economic Life." Trans. W.F. Stolper. *The Review of Economic Statistics*, vol. 17, no. 6 (November), pp. 105–15.

Latour, Bruno. 1987. *Science in Action: How to Follow Scientists and Engineers Through Society*. Milton Keynes, England: Open University Press.

MacKenzie, Donald, and Wajcman, Judy, eds. 1985. *The Social Shaping of Technology: How the Refrigerator Got its Hum*. Milton Keynes, England: Open University Press.

Mumford, Lewis. 1967, 1970. *The Myth of the Machine*, vol. 1: *Technics and Human Development*, vol. 2: *The Pentagon of Power*. New York: Harcourt Brace Jovanovich.

Noble, David. 1984. *Forces of Production: A Social History of Industrial Automation*. New York: Knopf.

Pinch, Trevor J., and Bijker, Wiebe E. 1984. "The Social Construction of Facts and Artifacts, or How the Sociology of Science and the Sociology of Technology Might Benefit from Each Other." *Social Studies of Science*, vol. 14, no. 3 (August), pp. 399–441.

———. 1987. "The Social Construction of Facts and Artifacts, or How the Sociology of Science and the Sociology of Technology Might Benefit Each Other." Pp. 17–50 in Bijker et al., 1987.

Price, Derek J. DeSolla. 1963. *Little Science, Big Science.* New York: Columbia University Press.

Schumpeter, Joseph. 1939. *Business Cycles.* New York: McGraw-Hill.

Seal, Lord Privy. 1971. *A Framework for Government Research and Development.* London: H.M.S.O. (contains a report by Lord Rothschild on "The Organization and Management of Government R&D").

Solow, Robert M. 1956. "A Contribution to the Theory of Economic Growth." *Quarterly Journal of Economics,* vol. 70, no. 1 (February), pp. 65–94.

Staudenmaier, John. 1986. *Technology's Storytellers: Reweaving the Human Fabric.* Cambridge, MA: MIT Press.

Winner, Langdon. 1986. *The Whale and the Reactor: A Search for Limits in an Age of High Technology.* Chicago: University of Chicago Press.

———. 1993. "Upon Opening the Black Box and Finding it Empty: Social Constructivism and the Philosophy of Technology." *Science, Technology, and Human Values,* vol. 18, no. 3 (summer), pp. 372–78.

4

Making Disciplines Disappear in STS

SUSAN E. COZZENS

One of the hallmarks of STS has been its interdisciplinary approach to questions of science and technology. Initially, however, most STS scholars and practitioners are trained in the standard disciplines with their associated methodologies—most frequently the history, sociology, and philosophy of science or technology. It is the resulting tension between the essential interdisciplinarity of STS "problems" and the powerful divisions of the disciplines that Susan Cozzens addresses in her essay. It is her contention that not until STS respondents—policy analysts, teachers, and researchers—truly transcend their disciplinarities will there be a central core of "STS Thought," of questions and concepts, that unifies the field.

Cozzens was from 1988 to 1993 the editor of *Science, Technology, & Human Values,* the official publication of the Society for Social Studies of Science and one of the most influential journals in the STS field. She is also a past director of the Office of Policy Support at the National Science Foundation. She has taught at the interdisciplinary Department of Science and Technology Studies at Rensselaer Polytechnic Institute and currently is Chair of the School of Public Policy at the Georgia Institute of Technology. As a result of her various positions, Cozzens has had the opportunity both to gauge the structure of STS as a field and to assist in remapping its contours. Among her publications is the co-edited volume of essays entitled *Theories of Science in Society* (1989).

As we read her essay, we should be thinking of how Cozzens'

call to rise above disciplinarity might affect students, first in terms of courses to be taken, then as we consider moving out into the workforce. It may also be helpful to compare Cozzens' depiction of the STS field with that of Fuglsang's more pluralistic view.

In 1990, I was asked to give a talk on the relationship between research, teaching, and practice in studies of science, technology, and society (STS). My specific assignment was to represent the "disciplinary perspectives" of the social sciences. But from the beginning, I rebelled against the task, and instead presented a "postdisciplinary perspective," because that was the viewpoint I believe we have to take if we are going to strengthen the research-practice relationship.

Because of its origins, my argument here has a rather personal tone, and draws very much on my own experience in the field. From the vantage point of nearly a decade later, I can report that the disciplines turn out to be just as strong as I reported in 1990, and that therefore the interdisciplinary integration I call for is still in need of advocates. I hope that my readers join the cause.

I begin by describing what STSers are about, by drawing an unconventional mental map of STS as a whole. Only after we have that map in mind will I turn to the tiny portion of it that reflects the academic disciplines, including the social science-based ones, to describe the situation there, complete with barriers and bridges among different kinds of STS analysis. In the end, I return to my postdisciplinary theme to describe the challenge of making disciplines disappear in STS.

The Map of STS

In 1988, I listened to a discussion among a set of academics who had been asked to choose priorities for future directions in the field of STS. In true academic form, instead of answering the question, they debated its terms. Was STS a discipline? (No one in this group said yes.) Was it a field? (There were dissenters even from this view.) Was it perhaps an "area," as in "area studies"? (Some took this position.) The discussion continued to puzzle me, until I attended my first meeting of the National Association for Science,

Technology, and Society (NASTS). Here was a set of people with more than academic commitments: they were teaching, they were writing, and they were out changing the world.

I remember standing in the lunch line with a person who had driven at the speed limit (not above!) all the way to the meeting because of his commitment to fuel conservation. I also remember a political figure asking how many in his audience had voted in the last national election. He was surprised when nearly everyone raised a hand—many more than his typical audience. It was at this meeting that I realized where the 1988 discussion group had gone wrong. They were looking for STS among academic descriptors. But what actually goes on under this label is much broader than any academic endeavor. STS is not a discipline, field, or area: it is a movement.[1]

The impetus for that movement is STS, The Problem. I do not need to describe that problem in any detail. We all know that science and technology are in society, and that they do not sit comfortably there. Sometimes they appear as the threats that come from larger-than-life weapons and environmental degradation. Sometimes they appear as salvation, through medical science, agricultural revolutions, or reproductive technologies. To some, their change-potential appears as investments, as choices among the technologies that will make or break businesses. Those choices in turn lead to changes for the rest of us that add up to new lifestyles, at work and at home. Through such routes, science and technology have become elements in most of the critical issues facing humanity: issues of peace and war, the environment, world health, universal subsistence. The map of STS, The Problem, must thus be very inclusive. It stretches around the world, from developed to developing countries, from high- to low-technology industry, from office and boardroom to factory and family room. (One can begin to follow this "big picture" in Figure 4.1.)

STS, The Problem, draws forth STS, The Response. The group of people involved in that response goes well beyond the set that would form up under an STS banner at a rally, or put STS bumper stickers on their cars. STS, The Response, includes industrialists, policy-makers, and members of public interest groups, all locked in debate and struggle over specific technological futures. It includes professional organizations of scientists and engineers who implicitly choose between passive acceptance of conditions set by others and the alternative of controlling their conditions of work and thus the intellectual and physical products they create. It also includes

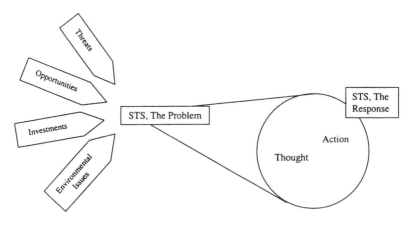

Figure 4.1
The Big Picture

academic observers—teachers and teacher-researchers—who examine and analyze STS, The Problem, in their everyday work.

Everyone who is part of STS, The Response, is involved in both thought and action in relation to science and technology in society, although the mix between the two differs in different positions. Industrial managers, for instance, are more concerned with doing than with understanding, although in order to do, they must have knowledge of their environments and a strategic view of how to respond to them. Likewise, those active in public interest groups are very strongly involved in thinking through problems in creative ways, but for a purpose, in order to change practice in some specific area. Academics, in contrast, are involved first and foremost in thought, through organizing materials for students and producing research. But then again, since our students go out into the nonacademic world affected by what we have shared with them, teaching is a form of action; and many academic researchers consult for action organizations using the knowledge they acquire in their academic activities. Clearly, no clean separation of thought and action, thinkers and doers, could be sustained in a concrete description of STS.

There are, however, three kinds of jobs in STS, The Response, that focus on thought: policy analysis, teaching, and research. Most policy analysts work for government or for organizations such as lobbies that exist to influence government; some work in academe. They are primarily information gatherers, evaluators, synthesizers, and extenders. These functions are very similar to the basic tasks involved in teaching and in research in schools, colleges, and universities. Policy analysts, teachers, and researchers are all of them thought specialists.

When we see these thought specialists as part of the response to STS, The Problem, it is immediately evident that their work will be improved if they interact, both with each other and with action specialists. If they do not, then their intellectual products stand in danger of being isolated, fragmented, and weak. It is all too easy, on the one hand, for policy analysts to stick close to their particular policy problems and not see the larger patterns of which they are a part. On the other hand, it is easy for academics to stay too far up in the conceptual clouds and never understand the application of their theories to any particular human situation. In contrast, when these thinkers see themselves and their problems as a part of the whole, their specific thought benefits from a growing body of interconnected knowledge addressing STS, The Problem. I call that body of knowledge STS Thought.

STS Thought—as an integrated whole, firmly anchored in STS, The Response to STS, The Problem—is an ideal, not a reality. Perhaps it is an ideal that is impossible to achieve. But it is nonetheless an ideal that gives us direction, places us within a context, and helps us choose among options for current action.

The Interdisciplinary Network

Of all the people involved in creating STS Thought, academic researchers are the easiest to find because they leave a paper trail. That is, conventionally they publish their thoughts. In concept, this practice should make their work more accessible to the others involved in STS, The Response. In practice, since articles must be approved for publication by specialists, who use esoteric language to set themselves apart from nonspecialists, publication often assures that this portion of STS Thought is almost entirely inaccessible to more than a small handful of persons.

Nonetheless, if we could crack the code of academic language, we would find that a large number of academic researchers have given some attention to STS, The Problem. I will refer to what these scholars do as "science and technology studies" (S&TS). The map of S&TS, it turns out, is quite easy to draw. The group of academics I mentioned earlier, who could not decide whether STS was a discipline, field or area, agreed quite readily to a map like the one in Figure 4.2.

On one side, we have the fragments of traditional disciplines that study science from those disciplinary perspectives. The big three are history, philosophy, and sociology of science. History and philosophy of science visibly split off from their parent disciplines several decades ago, and maintain only weak ties. Sociology of science, which originally focused on the institutional structure of science (Barber, 1962; Merton, 1973), later on small group and laboratory interaction (Crane, 1972; Price, 1963), and most recently on knowledge production (Clarke,1990; Fujimura, 1992; Knorr-Cetina and Mulkay, 1983; Star, 1991), has not done so. Recently, the study of science in literature and the scientific literature itself have emerged within this mix, from the ranks of rhetoric and English departments (Bazerman, 1988); and anthropology has become prominent in the 1990s (Hess and Layne, 1992; Forsythe, 1993; Traweek, 1988). As late as the 1980s, the hallmark of all these areas was their exclusive focus on science, not technology, with an accompanying limitation to science done in academic settings.

On the other side we have the students of technology: their research setting is industry. Traditionally, this group has been drawn primarily from economics and management (Mansfield, 1993; Nelson and Romer, 1996), but in recent decades a lively contingent of historians, political scientists, and in the risk area even philosophers, have been added (Hughes, 1983; Morone and Woodhouse, 1989; Weil and Snapper, 1989). Alongside the long-standing emphasis in this group on the production of technology, we have also seen the growth of research on the impact of technology in the workplace (Whalley, 1986; Noble, 1984; Zuboff, 1988). Sociologists

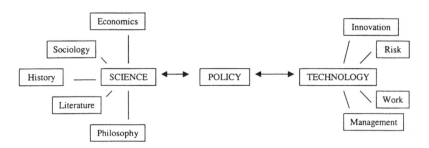

Figure 4.2
The Discipline Map

have joined the effort here, but quite different sociologists, by and large, from the ones who were studying science. New interdisciplinary interactions among historians of technology, sociologists of technology, and economists of R&D (research and development) have also emerged (Bijker, Hughes, and Pinch, 1987).

Betwixt and between the two are policy studies, the rare spot where some analysts are exploring the relationships between the institutions of science and those of technology. This bridging task is doubly hard. Not only does the science/technology split characterize academic research; it has also been maintained in government circles and thus characterizes policy itself. But in the 1980s, as the pace of university/industry interaction increased and as we saw the emergence of a range of policies to tie science to technology, this distinction began to disappear. A few hardy souls in the research community have moved into the intellectual space thus created. My own work, for instance, is on the power of scientific knowledge, which I see as the link between its external constituencies and its content (Cozzens, 1989). Others are trying to understand the implications of industrial sponsorship for university research (Peters, 1987). Still others are exploring connections between the nature of scientific expertise and how it functions in the political system (Nelkin, 1992; Richards, 1991; Martin, 1991; Jasanoff, 1990).

What is the relationship of traditional disciplines to this map, or more properly, to the people who are creating the map on a continuous basis through their research choices? Implicit in my description has been the point that, while disciplines are still very much in evidence in this research community, there is no one-to-one correspondence between the topics studied and the traditional disciplines. Historians appear in every region. Sociologists appear in every region. Philosophers appear in every region. Political scientists appear in every region.

Nonetheless, the disciplines continue to play a strong role in STS. Most people who study science and technology still do so within a single discipline. The History of Science Society, the Society for the History of Technology, and the Philosophy of Science Association are strong organizations, meeting regularly and providing career paths for many people. Similarly, there are sections of the American Political Science Association and American Sociological Association on science and technology, with hundreds of members who do all their work within those disciplines and who do not get involved in much interdisciplinary interaction.

Even among the minority who actively pursue interdisciplinary interaction and research in STS, disciplinary identities are strong. The STS interdisciplinarians tend not to discard their disciplinary training, but rather bring it to the forum and make it a positive contribution. In particular, disciplinary research methods and styles of problem choice seem to be cherished, even among those who have embraced postdisciplinary problem areas and concepts.

Finally, interdisciplinary interaction in science and technology studies tends to be pair-wise. Limited, specific areas of interdisciplinary exchange appear, rather than an across-the-board disciplinary mix. The new alliance I mentioned earlier among economists of R&D and historians and sociologists of technology on the nature of technological innovation is an example. Economists and sociologists, but few members of other disciplines, have hooked up recently in analyzing human resource questions in science and engineering. Similarly, political scientists, philosophers, and students of communication are interacting in studies of risk communication.

What results is not so much a melting pot as a multidisciplinary network, a very real structure of interaction that provides real opportunities for ideas to be exchanged but that consists entirely of smaller exchanges, linked together loosely. What is missing from this decentralized network is a core, a central set of questions or concepts that pull everyone together. I like to think that an opportunity for such a core to emerge is provided by the existence of the very interdisciplinary Society for Social Studies of Science (4S) and its journal, *Science, Technology, & Human Values*. But I must admit that the core has not appeared yet, even in 4S.

The network, however, seems to be getting stronger all the time. New links are forged, and new groups are brought in. In the future, the key to finding a core may lie in the generation of students just beginning to emerge from the interdisciplinary programs in STS. Some of these programs train students in bits of each of the constituent disciplines, then let them integrate the bits on their own. Others have tried to develop their own core of postdisciplinary courses, which is difficult given that the interdisciplinarians still have their disciplinary identities to which to cling. Under either model, the students emerging from these programs are less bound by disciplines than their teachers were and are more prepared to produce postdisciplinary research. A particularly promising sign is that many of these students have strong backgrounds in science or engineering, and thus may be able to bridge more than the social science/humanities gap.

Theory and Practice

Where do the intellectual products of the interdisciplinary network fit into STS Thought? Does interdisciplinarity bring us closer to the ideal? In one sense, yes, it must. Interdisciplinarity means integration of fragmented knowledge bases, and that is a significant part of the ideal of STS Thought. There is still a long way to go, however. This becomes apparent if we examine the connections between interdisciplinarity STS research and the other places where STS Thought is produced. If we put all the interdisciplinarians in one box, and look at the connections between that box and government, the public sector, and industry, the first thing we notice is that the network of connections outside the box is much sparser than the network within. There are fewer links, and the links that exist are weaker and more problematic for those involved.

The thickest area of the external network is probably with policy analysis. Ideas do flow in and out of government, and people move back and forth as well between academic positions and government jobs having to do with science and technology. The difficulties in achieving those two forms of movement should not be underestimated, however. Academics and policy analysts do not naturally speak the same language, and must put serious effort into translating into each others' terms if they are going to benefit from contact. Neither of their organizational homes consistently supports that effort. Furthermore, as I can attest from experience, while it is relatively easy to move from an academic position to a government one, it is exceedingly difficult to move in the other direction, because academic reward systems seldom respect and always downgrade achievements in other spheres. Similarly, ideas and people flow back and forth between public interest groups and the university, and between industry and the university, but, similarly, there are barriers that limit the flow to a relatively small phenomenon.

Finally, in considering the role of interdisciplinary STS research in STS Thought, we must consider its links to teaching. Since most academic researchers are also teachers, the barriers in this direction should not be as serious as those faced in relating to industry, public interest groups, and government. But there are barriers nonetheless.

First, there is the barrier that stands in the way of a person who is primarily an STS teacher but gets the urge to produce a paper trail—that is, to publish in a scholarly journal. Teaching STS is a generalist activity, and doing STS research is a specialist activity. The teacher's product is thus much more likely to be a generalist paper than a specialist one, and therefore may have trouble in a specialist journal review process. Scholarly journals insist on original contributions to an existing literature, but generalists may have trouble knowing enough of the literature to know what is new and what is not. Trying to get ideas to move in the other direction, however, may be just as difficult. STS generalists who try to read the specialist literature are confronted with the barriers of language and even problem definition described earlier. Adding insult to injury, many STS teachers are scientists and engineers, and therefore not trained in the methods of inquiry of the social sciences and humanities that characterize most of the published literature.

STS Thought

In short, the ideal of STS Thought—an integrated and accessible body of knowledge to inform STS, The Response—has not yet arrived. Science and technology studies are still too discipline-bound and cut off from the world of practice to serve as models. Policy may in the end be the best meeting ground for science and technology, thought and action: but at present, the meeting has barely begun.

I began this article, however, not with a problem but with a solution, or at least the choice of one. The solution lies in our determination to keep the whole map of STS, The Problem, in mind as the context for our work, and to see ourselves, not as sociologists or philosophers, teachers, lobbyists, researchers, or policy-makers, but rather as part of STS, The Response. If we see this context, then we will be better able to achieve STS Thought and eventually make the disciplines, and even the contradiction between thought and action in STS, disappear.

Concretely, I see a number of steps that my fellow researchers and I can take in this direction. One is to try to move away from specialization, toward broader relevance. We should be prepared to explain the importance of our work in terms that go beyond a specialist audience, and we should develop the language and skills to explain ourselves to a broader audience. When I was editor of *Science, Technology, & Human Values*, I tried to encourage these qualities in published articles, but I did not always succeed, and the pressures to publish specialist work were very strong. In any case, one journal has only a limited influence on general patterns. The determination to speak to a broader audience must be widely shared and consistently encouraged if the character of research in science and technology studies is going to change.

A second objective for the research community is to build its common core of problems and concepts. This involves paying attention to what we have in common, not just what separates us. For this task, interdisciplinary professional meetings are very important, since it is easier for us to speak past our differences and understand each other in person than in writing. Teachers and policy analysts can help researchers a great deal by recognizing similarities to which we have become blind ourselves. In addition, as I mentioned earlier, young scholars with interdisciplinary degrees will have an important role to play in developing a common core of concepts. The older generation of interdisciplinarians were trained to look for the differences, but the younger generation has survived by seeking the core. In the next few years they will have much to teach the rest of us.

Today's college students have a special role to play. Students are scattered across the big picture map in their future jobs. Every job they hold will put pressure on them to narrow their view of science and technology in society. Most commonly, they will be expected to see only one side of an issue—that of their firm, their government agency, or their part of society. Also frequently, they will be asked to set aside their values and their respect for all the people in the big picture, as though values and respect were unimportant. But once they have recognized STS, The Problem, and become part of STS, The Response, they are free to move beyond those pressures and represent a broader and deeper perspective in their thought and action.

In conclusion, while the ideal of STS Thought may still be far in the distance, there are steps we can take now that will move us in the right direction. Every movement has to be built, contribution by contribution, campaign by campaign. The STS movement is no exception.[2]

Notes

1. I later addressed the question in Cozzens (1993). Noting one root of STS in the counterculture movement of the 1960s, I raised some questions. "Where is the root of STS in the civil rights movement? Was it there

and withered? Was there a contradiction between the antimaterialist goals of the 1960s counterculture and the demands of the civil rights movement for equal access to a decent standard of living? Do the goals of STS express the problems of middle-class lives in the dominant culture? . . . Is there something bourgeois about STS as we, its current practitioners, have defined it? Is there something that makes STS particularly unattractive to those who have been excluded from the benefits of technological society?"

2. This essay was originally delivered as a talk at the 1990 annual meeting of the National Association for Science, Technology, and Society. An earlier version was published in the *Bulletin of Science, Technology, and Society*, vol. 10, no. 1 (1990), pp. 1–5, and is reprinted here in revised form with permission of that journal, currently published by Sage Publications.

References

Barber, Bernard. 1962. *Science and the Social Order*. New York: Collier.

Bazerman, Charles. 1988. *Shaping Written Knowledge: The Genre and Activity of the Experimental Article in Science*. Madison: University of Wisconsin Press.

Bijker, Wiebe, Hughes, Thomas, and Pinch, Trevor, eds. 1987. *The Social Construction of Technological Systems: New Directions in the Sociology and History of Technology*. Cambridge: MIT Press.

Clarke, Adele. 1990. "A Social Worlds Research Adventure: The Case of Reproductive Science." Pp. 15–42 in Susan E. Cozzens and Thomas F. Gieryn, eds., *Theories of Science in Society*. Bloomington: Indiana University Press.

Cozzens, Susan E. 1989. "Autonomy and Power in Science." Pp. 164–84 in Susan E. Cozzens and Thomas F. Gieryn, eds., *Theories of Science in Society*. Bloomington: Indiana University Press.

———. 1993. "Whose Movement? STS and Social Justice." *Science, Technology, & Human Values*, vol. 18, no. 3 (summer), pp. 275–77.

Crane, Diana. 1972. *Invisible Colleges*. Chicago: University of Chicago Press.

Forsythe, Diana. 1993. "The Construction of Work in Artificial Intelligence." *Science, Technology, & Human Values*, vol. 18, no. 4 (fall), pp. 460–79.

Fujimura, Joan H. 1992. "Crafting Science: Standardized Packages, Boundary Objects, and 'Translation'." Pp. 168–211 in Andrew Pickering, ed., *Science As Practice and Culture*. Chicago: University of Chicago Press.

Hess, David, and Layne, Linda, eds. 1992. *Knowledge and Society*, vol. 9: *The Anthropology of Science and Technology*. Greenwich, CT: JAI Press.

Hughes, Thomas. 1983. *Networks of Power: Electrification in Western Society, 1880–1930*. Baltimore: Johns Hopkins University Press.

Jasanoff, Sheila. 1990. *The Fifth Branch: Science Advisers as Policymakers*. Cambridge, MA: Harvard University Press.

Knorr-Cetina, Karin, and Mulkay, Michael, eds. 1983. *Science Observed: Perspectives on the Social Study of Science*. London: Sage.

Mansfield, Edwin, ed. 1993. *The Economics of Technological Change*. Aldershot, England: Elgar.

Martin, Brian. 1991. *Scientific Knowledge in Controversy: The Social Dynamics of the Fluoridation Debate*. Albany: SUNY Press.

Merton, Robert K. 1973. *The Sociology of Science: Theoretical and Empirical Investigations*. Ed. Norman W. Storer. Chicago: University of Chicago Press.

Morone, Joseph, and Woodhouse, Edward J. 1989. *The Demise of Nuclear Energy: Lessons for Democratic Control of Technology*. New Haven, CT: Yale University Press.

Nelkin, Dorothy, ed. 1992. *Controversies: Politics of Technical Decisions*. 3rd edition. Newbury Park, CA: Sage.

Nelson, Richard and Romer, Paul. 1996. "Science, Economic Growth, and Public Policy." Pp. 49–74 in Claude Barfield and Bruce L. R. Smith, eds., *Technology, R&D, and the Economy*. Washington, DC: Brookings Institution and American Enterprise Institute.

Noble, David F. 1984. *Forces of Production: A Social History of Industrial Automation*. New York: Knopf.

Peters, Lois S. 1987. *Academic Crossroads: The U.S. Experience*. Troy, NY: Rensselaer Polytechnic Institute.

Price, Derek J. DeSolla. 1963. *Little Science, Big Science*. New York: Columbia University Press.

Richards, Evelleen. 1991. *Vitamin C and Cancer: Medicine or Politics?* London: Macmillan.

Star, Susan Leigh. 1991. "Power, Technologies, and the Phenomenology of Conventions: On Being Allergic to Onions." Pp. 26–56 in John Law, ed., *A Sociology of Monsters: Essays on Power, Technology, and Domination*. London: Routledge.

Traweek, Sharon. 1988. *Beamtimes and Lifetimes: The World of High Energy Physicists*. Cambridge, MA: Harvard University Press.

Weil, Vivian, and Snapper, J. D., eds. 1989. *Owning Scientific and Technical Information*. New Brunswick, NJ: Rutgers University Press.

Whalley, Peter. 1986. *The Social Production of Technical Work: The Case of British Engineers*. Albany: SUNY Press.

Zuboff, Shoshana. 1988. *In the Age of the Smart Machine: The Future Work and Power*. New York: Basic Books.

II

Applications

5

An STS Perspective on Technology and Work

RUDI VOLTI

Because of its frequent problem-oriented nature, STS often leads scholars to focus on particular themes or issues. Such treatments are at once valuable with regard to the question at hand, but are also often revealing of broader STS conceptualizations. Work, as related to technological change—which is the theme of Rudi Volti's essay—is just such an issue. Here Volti critically examines the assumption that technological advance automatically results in unemployment, noting that it often creates far more *jobs* than it destroys. Yet he warns us that job loss/creation is not the whole story, because it is the nature of the *work*—and the wages earned thereby—that may ultimately be more important to the individual laborer. The introduction of new technologies has brought many changes to the workplace, but they have not done so as autonomous agents. In this reading Volti notes how technological changes have interacted with other forces to alter the nature of work, and in so doing he applies two concepts that are central to STS: *choice* and *context*.

Volti is professor of sociology at Pitzer College in Claremont, California, where he also teaches numerous STS courses. He is the author of the widely used STS textbook, *Society and Technological Change* (1995) and the recently published *Encyclopedia of STS* (1999). We may also note that Volti is personally fascinated by both historic and contemporary technologies of transportation—

especially cars, railroads, and motorcycles. Of the last named, he
currently owns four, each of which suggests why different technologies develop in the ways they do and reveals the different societal contexts in which they have become embedded, lessons trans-
ferable to the broader realm of STS studies.

In reading Volti's essay, we should ask ourselves how technology has affected work in our own experience, either as a student
or in full-time employment. In what ways might such technological
changes as computerization and electronic communications affect
that work experience in the future? Who will those changes affect
most directly, and will they likely be for the better or the worse?

In 1982 I bought my first computer. Costing $225 (about $600
in late 1990s dollars), it had four kilobytes of memory and was
capable of doing little more than playing some simple games. Two
years later, I took a big step forward: for $2,000 I acquired a com-
puter with 64 kilobytes of memory and an accompanying software
package. This was a more serious machine, and with it I eventually
wrote a manuscript. Still, it was slow, had a tiny monochrome display, and required the use of a host of complicated codes for its operation. Today, for less than half the price of my old computer one gets
a machine with a memory that is measured in megabytes, a vastly
faster processing speed, a color display, and a number of features
that were confined to R&D laboratories in the early 1980s.

The story of the personal computer can be repeated in many
other technological realms. In 1939 Pan American Airlines initiated
the first scheduled air service across the Atlantic. Its Boeing flying
boats had a top speed of 180 mph and carried seventy-four passengers, each of whom paid $675 (about $8,000 at today's prices) for a
round-trip ticket. Today a Boeing 747 jumbo jet transports up to
500 passengers at a speed of nearly 600 mph, and costs the pas-
senger perhaps one-tenth the 1939 fare. In the mid-1950s, a color
TV of uncertain reliability delivered a marginally acceptable picture at a price that was beyond the means of most consumers;
today's high quality, inexpensive color televisions are universal
household items. Finally and most significantly, technological
advances such as antibiotics, computer-assisted tomography, organ
transplants, magnetic resonance imaging, kidney dialysis, artificial
joints, and heart pacemakers have made major contributions to the
reduction of death, disease, and discomfort.

Examples of technological progress can be multiplied almost indefinitely, making it seem as though advances in science and technology are synonymous with progress in general. Things are not so straightforward, however, and one of the tasks of science, technology, and society (STS) is to show the limitations of this view. This does not mean that STS is inherently negative in its approach to science and technology. To the contrary, through studies of the past and present, STS research has deepened our understanding of how scientists, engineers, and technicians have transformed the world.

STS is critical in the exact sense of the word: it strives to provide careful, comprehensive evaluations of the nature of scientific and technological change and the forces that impel it. To be sure, STS is not unique in taking a critical approach to science and technology. Efforts to come to grips with the consequences of science and technology are a prominent feature of our era. Often, the focus is on the unfortunate consequences of scientific and technological advance like environmental degradation, the proliferation of nuclear weapons, and an accelerated pace of life that seems overwhelming at times. STS attempts to go beyond criticisms of this sort by trying to understand *why* science and technology have developed the way they have. It does this by exploring how social structure, culture, politics, and economics have been intertwined with the development of science and technology.

Work and Technological Change

STS scholarship has covered everything from sixteenth-century scientific experiments with vacuum pumps to the invention of the zipper (Shapin and Schaffer, 1985; Friedel, 1994). In this section I will narrow my focus somewhat, and use an STS perspective to discern the interconnections between technology, work, and employment. This, of course, is a topic of more than academic interest. For most of us, gainful employment is not an option. Work is necessary for physical survival, and it often is a central component of our identity as individuals. It also happens to be an area of life that has been heavily affected by technological change, and—more indirectly—by the advance of science.

One common assumption about technological advance is that it results in unemployment. Here is a place where popular belief and scholarly judgment are on different tracks, for economists are virtually unanimous in their belief that technological advance does not result in an overall loss of jobs. This view seems to violate common sense; after all, if a new machine takes the place of twenty manual laborers, have not twenty jobs been lost? The mistake here is to equate particular *jobs* with *work* in general. Technological change may cause some employees to lose their jobs, but it does not reduce employment in aggregate.

A bit of analysis shows why this is so. Labor-saving technologies usually are implemented because they lower the cost of production. When production costs are lowered, several things are possible. First, the reduced cost of production allows producers to charge a lower price for their product. Consumers may therefore buy greater quantities of the product, resulting in more job opportunities for employees involved in sales, maintenance, administration, and the other tasks connected with the manufacture and distribution of the product. Alternatively, consumers may continue to buy the same quantity of the product, leaving them with more money to buy other goods and services. This will increase employment opportunities in the industries that produce these goods and services.

There is of course another possibility: the firm that replaced the twenty workers with the machine may be able to charge the same price as before, so the lower production costs will result in increased profits. But something will be done with the profits. They may go to the stockholders as earnings or they may be distributed as higher wages and salaries. In turn, some of this extra income will be spent, and the increased levels of spending will increase job opportunities in the firms that supply the things that are purchased. Alternatively, the manufacturers may plow the additional profits back into the firm by investing in more physical capital, thereby increasing job opportunities in the firms that supply the investment goods.

So far, so good; technological advance is acquitted of the charge of being a job-killer. But this is not the end of the matter. Although technological advance does not reduce overall levels of employment, it does change the mix of jobs, and in doing so, it may result in the redistribution of incomes. In our example, the employer of the workers who lost their jobs may have taken the increased profits and distributed them to the stockholders in the form of higher earnings. This increased the stockholders' spending power, and in so doing it created new job opportunities. However, stock ownership is not distributed evenly throughout society, and stock earnings disproportionately go to upper-income segments of the population. Consequently, most of the increased income went to relatively wealthy individuals who might have spent it on the services provided by portrait painters, travel agents, and exclusive boarding schools. On the other hand, if the increased income had gone for investment, the beneficiaries might have been engineers, systems analysts, and prototype machinists. And while all this has been going on, some of the twenty redundant manual laborers may still be collecting unemployment checks.

This of course is a hypothetical case. The important point is that technological advance may eliminate some jobs while stimulating the creation of others, and that the process creates both winners and losers. This has been readily apparent in recent years as the economy has continued to shift from the manufacture of things to the production of services. Although total output in the manufacturing sector has increased, the number of manufacturing employees has dropped by a small amount in absolute terms, and by a very large amount relative to total employment. In 1979 the manufacturing sector employed 23.4 percent of the labor force; by 1995 it provided jobs for only 15.8 percent (Mishel, Bernstein, and Schmitt, 1997, p. 186). At the same time, the percentage of jobs in the service sector (i.e., sales, finance, education, government, entertainment, health care—just about anything that does not produce a tangible product) went from 70.5 percent to 79.2 percent, while the primary sector (comprising such activities as farming, mining, and fishing) absorbed the remainder. Productivity-enhancing technological changes were not the only reasons for the relative decline of manufacturing employment and the rise of the service sector, but they were certainly a key element of the process.

The increased productivity of the manufacturing sector might be a cause for celebration, except that it has contributed to widening inequalities in the distribution of income in the American economy. Although some service-sector occupations such as physician and attorney offer high rates of compensation, for the most part, service occupations are not as well rewarded as those in the manufacturing sector. In 1989, total compensation (wages and salaries plus benefits) averaged $15.60 in the manufacturing sector versus $11.96 in the service sector, a difference of 30.4 percent (Mishel and Frankel, 1991, p. 106). The expansion of the service sector has created an abundance of jobs in recent years, but many of them pay meager wages. More than 83 percent of the new jobs created from 1989 to 1995 occurred in health care, retail trade, and temporary services—

industries with a preponderance of low-wage jobs (Mishel, Bernstein, and Schmitt, 1997, p. 184).

While many workers are employed in low-skill, poorly rewarded jobs, technological advance has not generated large numbers of high-paying jobs for well-educated workers with advanced skills. According to one study, from 1986 to 1996, high-tech industries (as indicated by the proportion of their budgets devoted to research and development) added about 400,000 workers, a paltry 2.9 percent of the 14 million jobs created during this period (Luker and Lyons, 1997).

Changes in the distribution of occupations has been partly responsible for the widening income disparities that have been a feature of the American economy in recent years. The trend of greater income inequality can be seen by dividing households into five segments or *quintiles*, and observing what has happened to their relative shares from 1973 to 1996, expressed as a percentage of total income (adapted from Ryscavage, 1999, p. 59) (see Table 5.1).

For full-time male workers in the lower quintiles the drop has been even more precipitous than it has been for households. Between 1973 and 1992 the wages and income of the bottom quintile fell by 23 percent, while the increases of those of the fourth quintile fell by 21 percent. Workers in the third quintile saw their incomes fall by 15 percent, while the incomes of those in the fourth quintile experience a drop of 10 percent. All the income gains to male workers went to those in the top quintile, where incomes increased by 10 percent during this period (Thurow, 1996, p. 23).

Table 5.1
Income Distribution by Quintiles,
as Percentages of National Income

	1973	1996
first quintile (lowest)	4.2	3.7
second quintile	10.5	9.0
third quintile	17.1	15.1
fourth quintile	24.6	23.3
fifth quintile (highest)	43.6	49.0

To be sure, the widening income inequalities of recent years cannot be attributed solely to technological change. Many other factors have been involved, such as a decline in the purchasing power of the government-set minimum wage, increased immigration, the erosion of the power of labor unions, and changes in the size and composition of households. Changes in income distribution also can be linked to the globalization of the American economy. Here, however, technological advance has played an important role in facilitating the movement of goods and services across national boundaries. Some of these developments, like the widespread use of e-mail and fax machines, are closely linked to the rapid advance of integrated circuits, microprocessors, fiber optics, and other elements of high technology that have emerged in the last two decades. These electronic technologies make it possible to communicate across the globe almost instantaneously, making it easy for an investor in Frankfurt to buy shares in an American company listed on the New York Stock Exchange, or for an Osaka-based firm to stay in close contact with the manager of its factory in Thailand.

Other technological innovations are not as glamorous as high-tech communications, but they have made significant contributions to globalization. One of the most important of these is the shipping container, a large metal box that allows a great variety of products to be carried by ship, train, and truck without incurring the expense, damage, and pilferage that accompanies the unloading and reloading of individual items. The simple container is one of the reasons that a stereo set manufactured in Malaysia can be sold in a store in Chicago at about the same price that it commands in the country where it was made.

The increased importation of foreign goods has led to some job losses in the U.S. manufacturing sector, although economists do not agree on its extent (Freeman, 1995). As with the use of labor-saving technologies, the globalization of manufacture has not decreased employment overall because the expansion of the service sector has more than offset job losses in manufacturing. But globalized manufacturing has made some contribution to widening income disparities by diminishing the number of well-paying manufacturing jobs in the United States. According to one estimate, the increased importation of manufactured goods resulted in a 5.9 percent decline in manufacturing employment between 1978 and 1990 (Mishel, Bernstein, and Schmitt, 1997, p. 192). Once again, manufacturing workers with skills not easily transferred to other industries may be victimized by technological change, although the process has not been a simple case of machines displacing human labor.

In addition to altering distribution of incomes, technological advances have profoundly changed the nature of work. This is not a new development; on many occasions the emergence of new technologies has strongly affected working lives. Indeed, nothing in the history of humankind has been as significant as the shift from gathering and hunting to sedentary agriculture, a process that began perhaps 8,000 years ago (Rindos, 1984). At first, an agricultural way of life and work entailed the cultivation of domesticated plants through the use of a few simple tools. The subsequent development of agricultural technologies, especially the construction of irrigation systems, greatly expanded food supplies that fed much larger populations while providing the foundation for what we call civilization. But at the same time, irrigated agriculture required more toil. Compared to gathers and hunters, agriculturists have had to work much longer hours at harder, more monotonous tasks (Lee, 1968). Agriculture fed large numbers of people and served as the foundation of sophisticated cultures, but it required the unremitting efforts of the vast majority of the population who labored as peasant farmers.

The other great era of change in human history, the industrial revolution (the first phase of which began around the mid-eighteenth century), was closely tied to a great outpouring of new mechanical technologies—the textile machinery, steam engines, and blast furnaces that transformed economies around the globe. These technologies were responsible for a great increase in the output of manufactured goods, and people's living standards improved accordingly. At the same time, however, for many workers, technological advance meant toiling in a factory for up to seventy hours per week. Many industrial jobs were monotonous in the extreme, for mass-produced industrial goods depended on the use of automatic machinery and standardized procedures that required little in the way of skill. The culmination of this trend was the assembly line set up by the Ford Motor Co. early in the twentieth century. The efficiencies brought about by the assembly line made automobiles affordable for the multitude, but at a cost of numbingly repetitive low-skill work and the surrender of all control to the dictates of "the line" (Lichtenstein and Meyer, 1989).

Choosing Technologies

The brief review presented above has highlighted how technological changes have influenced work and employment. But to leave the story at this point is to miss one of the essential insights of STS, that *technology does not produce social change as an independent agent*. Technology can be a powerful impetus for change, but it always works in conjunction with political, social, cultural, and economic forces. To make technology an independent actor in affecting social change is to subscribe to "technological determinism," the belief that technology moves according to its own inner dynamic, unaffected by social arrangements, culture, and even human volition (for a discussion of technological determinism, see Smith and Marx, 1994).

One of the major contributions of STS has been the development of a more sophisticated understanding of the causes and effects of technological change. From an STS perspective, technologies are purposely developed and chosen by particular actors; consequently, the needs, desires, and intentions of individuals and groups are of surpassing importance in determining the kind of technologies that are created and the uses to which they are put. The assembly line transformed many manufacturing operations because it increased worker productivity; in this sense it was a superior technology in comparison with other ways of making cars. It vaulted Ford Motor Co. to the top rank of automobile manufacturers, and it made Henry Ford one of the richest men on the planet. Some of the productivity gains were passed on to the line workers, whose wages were among the highest in the manufacturing sector. But these high wages came at a great cost in terms of physical and mental fatigue, and many workers were incapable of keeping up with the demands of the line for more than a few months.

Given these tradeoffs, it is anything but certain that a managerial system with significant worker input would have adopted the assembly line. And if an assembly line had been installed, it is likely that it would have been designed with workers' needs in mind, even if this resulted in lower productivity. The former was what happened in Britain, where strong labor unions were able to delay the installation of assembly lines in many automobile firms until the 1930s. Productivity in the British automobile industry was lower, and hence wages were also lower, but for many workers the retention of some control over the pace of work made this a reasonable tradeoff (Lewchuk, 1989).

Choices also may be significant even when the same technology is injected into different work settings. Consider the divergent purposes to which a business firm may put a computer network.

One firm may use the network to reduce labor costs by having their data processing done in a Third World country. In contrast, another firm may use its network to give its rank and file employees access to information and to supply them with decision-making tools like statistical analysis software. When this is done, the network will increase employee participation in the firm's operations and make for a "flatter," less hierarchical organizational structure. Once again, social and political relationships may determine how a technology is used; the residents of the middle and upper echelons of an organization's managerial staff may not be overenthusiastic about using computer network technology to empower lower level employees if it means a reduction in their own role.

The divergent uses to which a new technology may be put also can be seen in higher education. Far from being insulated from the workaday world, colleges and universities are subjected to many of the same pressures found in the profit-making sector, especially the desire to hold down costs and to increase productivity—which may of course be defined quite differently by students, professors, and administrators. In this environment, innovations like personal computers and the internet can be used for very different, even conflicting, purposes. On the one hand, electronic technologies can be used to enhance a student's educational experiences by making scholarly publications readily available, by facilitating interactive approaches to learning, and by putting a wealth of information at a student's disposal. On the other hand, the very same technologies can be the basis of "digital diploma mills" that offer little more than electronic correspondence courses taught by members of an academic proletariat that rarely interact with students and have little chance of gaining a secure academic position (see the articles that appear in *Thought and Action*, spring 1998).

Technological change has given work a remarkable fluidity; many of the jobs done today did not exist a generation ago, and it can be confidently asserted that many of the readers of this chapter will be engaged in work activities that are unknown today. Since work usually takes up about half of an employed adult's waking hours, and often lies at the core of his or her personal identity, technological changes on the job will have major consequences. Still, it bears repetition that technology is a powerful force, but it is not an autonomous force. Both science and technology are human creations, and as such they are driven by the choices that people make.

In the realm of science, researchers decide what problems to

pursue, funding agencies determine which projects will receive support, and the culture in general may exert some influence on how
scientists see and interpret the things they study. With technology
the significance of socially shaped choices is even more evident.
Technologies are chosen not just because they "work" according to
abstract technical standards. The determination of whether or not
a technology may be adjudged "successful" in many instances
depends on who is making the assessment, and where their interests lie.

Technology in Context

The exercise of power over the choice of technologies is one
aspect of a concept that has been central to STS scholarship: the
significance of the *contexts* in which science and technology operate.
Context, in the STS idiom, refers to the particular social, economic,
political, and cultural structures and processes that shape science
and technology. In regard to work, the technologies used on the job
reflect such contextual elements as worker skills, consumer prefer-
ences, managerial practices, government regulations, the price of
capital, and the strength of labor unions. Many of these individual
elements are closely connected to other contextual elements. For
example, an industry or a business that is able to attract a large
amount of capital investment is likely to employ capital-intensive
technologies that require workers with advanced technical skills. A
key task (and a source of intellectual pleasure) of STS research is
delineating the linkages between the various elements that collec-
tively constitute the contexts of different technologies.

My brief exploration into the relationship between work and
technology has stressed how technology operates within a social
context, a context that often reflects disparities of power and
authority. As noted above, this perspective makes STS a critical
enterprise in both the general and the specific sense of the word.
And yet, while STS takes a critical stance, it is also a hopeful
endeavor. From an STS point of view, science and technology are
not juggernauts that operate beyond the reach of human control.
The STS axiom that the development of science and technology is
guided by human choices means that we have the ability to shape
and control science and technology for our benefit. At the same
time, the knowledge that science and technology develop within
specific social contexts should alert us to the likelihood that real

world choices may have little to do with the common good. STS scholarship has shown us that science and technology are subject to many of the same forces that shape society in general. This gives us a reasonable hope that a better comprehension of these forces will enlarge our understanding of how science and technology develop, and how they may be brought into closer accord with human needs.

References

Freeman, Richard B. 1995. "Are Your Wages Set in Beijing?" *Journal of Economic Perspectives*, vol. 9, no. 3 (summer), pp. 15–32.

Friedel, Robert. 1994. *Zipper: An Exploration in Novelty*. New York: Norton.

Hacker, Andrew. 1997. *Money: Who Has How Much and Why*. New York: Simon and Schuster.

Lee, Richard B. 1968. "What Hunters Do for a Living, or How to Make Out on Scarce Resources." Pp. 30–48 in Richard B. Lee and Irven DeVore, *Man the Hunter*. Chicago: Aldine.

Lewchuk, Wayne. 1989. "Fordism and the Moving Assembly Line: The British and American Experience, 1895–1930." Pp. 17–41 in Lichtenstein and Meyer, *On the Line*.

Lichtenstein, Nelson, and Meyer, Stephen. 1989. *On the Line: Essays in the History of Auto Work*. Chicago: University of Illinois Press.

Luker, William J., Jr., and Lyons, Donald. 1997. "Employment Shifts in High Technology Industries, 1988–1996." *Monthly Labor Review*, vol. 120, no. 6 (June), pp. 12–25.

Mishel, Lawrence, Bernstein, Jared, and Schmitt, John. 1997. *The State of Working America, 1996–1997*. Armonk, NY: M.E. Sharpe.

Mishel, Lawrence, and Frankel, David M. 1991. *The State of Working America, 1990–1991*. Armonk, NY: M.E. Sharpe.

Rindos, David. 1984. *The Origins of Agriculture: An Ecological Perspective*. Orlando, FL: Academic Press.

Ryscavage, Paul. 1999. *Income Inequality in America: An Analysis of Trends*. Armonk, NY: M.E. Sharpe.

Shapin, Steven, and Schaffer, Simon. 1985. *Leviathan and the Air-Pump: Hobbes, Boyle, and the Experimental Life*. Princeton, NJ: Princeton University Press.

Smith, Merritt Roe, and Marx, Leo, eds. 1994. *Does Technology Drive History? The Dilemma of Technological Determinism.* Cambridge, MA: MIT Press.

Thought and Action: The NEA Higher Education Journal, spring 1998.

Thurow, Lester C. 1996. *The Future of Capitalism: How Today's Economic Forces Shape Tomorrow's World.* New York: William Morrow.

Volti, Rudi. 1995. *Society and Technological Change.* New York: St. Martin's Press.

———. 1999. *The Facts on File Encyclopedia of Science, Technology, and Society.* 3 vols. New York: Facts on File.

Wilk, Richard R. 1996. *Economies and Cultures: Foundations of Economic Anthropology.* Boulder, CO: Westview Press.

6

Science-Technology-Society
and Education:
A Focus on Learning
and How Persons Know

ROBERT E. YAGER

STS received much of its early impetus as an educational initiative at the college and university level. Beginning in the 1980s, however, there was also a strong movement within the science education field for adopting an STS approach to learning science and technology within the schools. Robert Yager, a strong proponent of STS as an approach to science and technology education, offers a brief history of the major National Science Foundation and National Science Teachers Association (NSTA) efforts at identifying the characteristics and goals of effective STS education. In such an approach "learning" becomes the major outcome, not through thinking about STS as a body of specific content to be mastered by students, but rather as a vehicle for such knowledge development.

Yager argues for a "constructivist" way of knowing, one in which learning is developed in the mind of each person, not transferred rotely from one person to another. Here it is important to note that Yager and other constructivist educators are using the terms somewhat differently than constructivist scholars such as Wiebe Bijker (see his essay in this volume), who view the creation of scientific knowledge and technological artifacts and systems primarily as the result of a socially mediated or "constructed" process. Indeed the learning of social constructivist doctrine may even be at odds with constructivist STS science learning.

Yager, a biologist by training, has himself been a key participant in the shift in thinking about science and technology education in the schools. He is professor of science education and head of the Science Education Program at the University of Iowa, where he is well known for promoting the NSTA Iowa Chautauqua Program, an in-service effort designed to introduce science teachers to the STS constructivist teaching approach in science education. He has served as president of the National Science Teachers Association, the National Association of Science, Technology, and Society, and the National Association for Research in Science Teaching, in which capacities he has been a tireless voice of advocacy for STS. He is as well the author and editor of numerous articles and books on science education, including *Science/Technology/Society as Reform in Science Education* (1996).

As a science educator Yager is personally most interested in effecting changes in the way the next generation of science teachers will approach their work. He sees changing teachers' roles and the shift in teacher education that will accompany them as the biggest challenge ahead, because based on his research, STS students will rise naturally to the learning opportunities afforded them in an increasingly scientific world. As we read his thoughts on STS teaching, we should consider ways in which we might contribute to our own constructivist learning both inside and outside the classroom.

Science-Technology-Society is a special version of the STS acronym proposed by John Ziman, a British physicist and science educator, to indicate emerging programs and courses that looked broadly at science by relating science to technology and both enterprises as based in the minds and actions of human beings (Ziman, 1980). More expansively, science is seen as the attempt of humans to explain the natural world in ways that permit evidence to be accumulated to the point that most accept the explanation as valid, with technology, as human manipulation of the natural world to produce devices to benefit the human condition. For obvious reasons STS has attracted much attention in education, especially among those interested in an education where the outcomes of learning are useful to the lives of learners.

History of STS in Education

Since recorded time there have been frequent calls for improving schools. One of the earliest took place over 2000 years ago when Aristotle was called to study the schools of ancient Athens. Educators have called it one of the earliest curriculum studies ever completed. Aristotle identified two different types of schools or justifications for education: schools that taught what was considered important and correct for the youth of the society, and those that taught students the information and skills that they could use in their own lives. He ended his report by concluding that he could not recommend either purpose over the other, for neither was right or wrong, and each had separate defenders. In a sense, then, society has debated the major function of formal education since it decided to organize schools. Is the fundamental purpose of an education a matter of giving people a body of knowledge that society says is important? Or is it to prepare them to live a better life in terms of skills and information?

This dual justification has been central to the educational enterprise in the United States. Prior to 1800 both high schools and higher education were rarely experienced by most people. Most who aspired to formal education were doing so to prepare for the clergy. In the early 1800s the common science courses in secondary schools were navigation, surveying, and agriculture. Many today would call such a curriculum STS-centered.

As education in high school and college increasingly became accessible to more people, and as formal science matured, discipline-based science courses emerged. Thus, by the late 1800s science in schools and colleges had become organized around what we now view as the standard disciplines. Most schools were swift in offering courses in physics after Harvard University listed it as an entrance requirement in 1892. Ten years later Harvard added chemistry as a requirement. Biology did not become a common course until the 1920s, and it is rarely required by course title. Most students complete biology, however, as the first high school course at the tenth grade level. In contrast, such interdisciplinary and STS fields as the earth sciences have never achieved the same prominence as physics, chemistry, and biology.

Some scholars view Jefferson as one of the first STS advocates in the United States when he urged that all persons must be taught to think and to make decisions for themselves—a necessity if a democracy is to work. Every major educational reform in U.S. history has been an attempt to define education in ways that it could and would affect the lives of learners in positive and planned ways.

STS as an educational reform became prominent in the 1970s with two curriculum efforts initiated in the United Kingdom. One was called Science in Society, a second Science in a Social Context. The work of Ziman in the latter prompted some of his writing about STS and popularization of the STS acronym there. STS as a science-education term was introduced in the U.S. when the U.K. projects became conversation pieces at national meetings and in 1978 when the National Science Foundation (NSF) funded Project Synthesis to determine where U.S. science education was and where it should go in the immediate future. STS became an established science education movement in the U.S. when the National Science Teachers Association proclaimed that the major purpose of science education was to produce persons who were scientifically and technologically literate (NSTA, 1990).

Project Synthesis itself pursued four basic goals, as summarized in Table 6.1. In searching for ways to achieve these goals Norris Harms, Project Synthesis director, looked at five organizational schemes, of which one was STS (Harms, 1977). Following this, NSTA in 1982 initiated its "Search for Excellence." In 1984 an attempt to identify STS programs across the United States which

Table 6.1
The Goals that Framed Project Synthesis

1. Science for meeting personal needs. Science education should prepare individuals to use science for improving their own lives and for coping with an increasingly technological world.

2. Science for resolving society issues. Science education should produce informed citizens prepared to deal responsibly with science-related societal issues.

3. Science for assisting with career choices. Science education should give all students an awareness of the nature and scope of a wide variety of science—and technology—related careers open to students of varying aptitudes and interests.

4. Science for preparing for further study. Science education should allow students who are likely to pursue science academically as well as professionally to acquire the academic knowledge appropriate for their needs.

had certain features were identified. The NSTA defined excellence in the area of science-technology-society as programs that accomplish the four actions indicated in Table 6.2, which are closely related to the four goals of Table 6.1.

The "Search for Excellence" further proclaimed that exemplary STS programs should include six opportunities:

1. To learn about the energy involved in a variety of areas—from taking long, hot showers, to potential indoor pollution resulting from sealing houses too tightly against drafts, to the world impact of increasingly rapid growth of energy use throughout the world.

2. To discuss natural control of populations, the effect of technologies on population growth, and the impact of rapid changes of population growth on specific subsets of the world society.

3. To develop student awareness of the effects of personal and societal decisions on all aspects of the environment—from paper and food on the floor of the cafeteria, to the balance of gases in the atmosphere, to the "noise" of home stereo systems.

4. To encourage students to question the apparent waste in various technological programs as well as the potential benefits.

5. To deal with the complexity of day-to-day decisions related to science and technology. For example, while it can be demonstrated that 45 mph is a more energy-effi-

Table 6.2
Features Characterizing Excellent STS Programs

1. Prepare individuals to use science for improving their own lives and for coping with an increasingly technological world.

2. Prepare students to deal responsibly with technology/society issues.

3. Identify a body of fundamental knowledge which students may need to master in order to deal intelligently with STS issues.

4. Provide students an accurate picture of the requirements and opportunities involved in the multitude of careers available in the STS area.

cient speed at which to drive most autos, the national speed limit is 65 mph. The sociology behind such regulations should be understood along with the technology. Similarly, the automation of supermarkets has been technologically feasible for many years. However, the sociology involved in gaining public acceptance for this system has slowed down its implementation.

6. To consider such issues as weather control, test tube babies, genetic engineering, space shuttles, nuclear energy, and a myriad of technological developments that require an education which enables individuals and groups to make intelligent decisions on support or opposition to such technologies.

In 1990 the NSTA published a position paper which seemed to suggest STS would be *the* reform movement of the coming decade. It described STS programs as those where certain features were in evidence. Table 6.3 is a listing of these features. The position paper further defined STS as the:

> teaching and learning of science/technology in the context of human experience. It represents an appropriate science education context for all learners. The emerging research is clear in illustrating that learning science in an STS context results in students with more sophisticated concept mastery and ability to use process skills. All students improve in terms of creativity skills, attitude toward science, use of science concepts, and process in their daily living and in responsible personal decision-making (NSTA, 1990–1991, p. 47).

Nonetheless, there was no clear progression to reform. Purists from the discipline-oriented sciences began anew by identifying the concepts and the skills that all should know. These were scientists who headed the curriculum projects from the 1960s, where the post-Sputnik reforms were centered around the position of Jerrold Zacharias, the MIT physicist who conceptualized the Physical Science Study Committee (PSSC) program. For Zacharias, if taught in the way it is known to scientists, science will be interesting and appropriate to all.

Bill Aldridge, the executive director of NSTA and a physicist, became an outspoken critic proclaiming "STS is not science!" In the heat of this criticism, NSF abruptly stopped funding projects with

Table 6.3
NSTA STS Program Features

1. Student identification of problems with local interest and impact.

2. The use of local resources (human and material) to locate information that can be used in problem resolution.

3. The active involvement of students in seeking information that can be applied to solve real-life problems.

4. The extension of learning beyond the class period, the classroom, the school.

5. A view of science content which is more than concepts which exist for students to master on tests.

6. An emphasis on process skills which students can use in resolving their own problems.

7. An emphasis upon career awareness—especially related to science and technology.

8. Opportunities for students to experience citizenship roles as they attempt to resolve issues they have identified.

9. Identification of ways that science and technology are likely to impact the future.

10. Some autonomy for students in the learning process (as individual issues are identified).

an STS focus. Many science educators involved with related reform efforts such as the AAAS Project 2061 distanced themselves from the STS rationale. The American Chemical Society halted classifying its innovative *Chemistry in the Community* (or *ChemCom*) textbook as an example of STS. Even the National Science Education Standards do not refer to STS. Instead language advanced to characterize STS was employed without use of the acronym per se.

Interestingly NSF avoidance of controversy and the halting of any course development, curriculum frameworks, and staff development activities did not stop the effort. Instead projects continued to proliferate, such as Berkeley's Science Education for Public Understanding Program (SEPUP), the Smithsonian Institution's Science and Technology for Children, Northwestern University's Materials World Modules, and the Montgomery County Events-Based Science. These examples continue to point toward a major reform in science education, with results that are more positive in students than those found with traditional approaches. STS science education solves many problems associated with underserved groups while also meeting the most recent goals for the field (Yager, 1993).

Broadening the Prospectus
of Science and Technology

A major feature within STS educational reform, but not one easily achieved, has been the inclusion of technology with science in the typical secondary or college program. Getting persons dedicated to the science disciplines interested, comfortable, informed, and disposed to consider technology together with science is difficult. In secondary schools technology is often relegated to the vocational areas and seen as relevant only for the noncollege-bound student. In liberal arts colleges technology as a discipline is seldom found.

Karen Zuga, a specialist in technology education, has written about this unfortunate split between science and technology. She reports, "that the communities of technology and science educators have been passing in our schools and universities as two ships pass silently in the night without speaking to each other about their relationship" (Zuga, 1996, p. 227). Scientists and technologists have grown apart due to their specializations, such that they are "unable to speak each other's language," much as C. P. Snow had noted about the different cultures of science and the humanities (Snow, 1959). In the educational arena technology and science are perpetuating this artificial separation of technology and science in the next generation, through the students they teach.

In the modern era technology and science have grown apart as each area has developed different modes of operating. Science became more abstract and theoretical, while technology became concrete and practical, separating theory from practice. The growing knowledge base and the complexity of each area forced scientists and technologists to specialize. Educators have mimicked this separation through subject matter distinctions that transfer the resulting abstract/concrete language dichotomies to new generations of students (Kowal, 1991). In the schools this has further resulted in science educators focusing their attentions upon the brightest college-bound students, while relegating technology for those students in the vocational areas.

Philosophers and educators such as John Dewey (1925) and Alfred North Whitehead (1925) long ago recognized the artificiality of such a separation of technology and science in education. It is in this tradition that Zuga concludes:

> It is time to rejoin technology and science. Technology and science educators are beginning to realize that the reunification is occurring at the forefront of investigation as new areas of inquiry such as biotechnology are created by researchers. Technology and science educators' task in classrooms is to search for ways in which to open and create a discussion that helps to integrate technology and science for students (Zuga, 1996, p. 237).

It is important to note that STS takes students and teachers out of the disciplines of science per se and moves the curriculum beyond the typical topics and specialized skills of both science and technology. This does not mean that particular science and technology concepts and skills are not important. Instead it is the context in which students encounter them that is critical. Teaching either concept or process skills outside of a real world context invites failure, especially when teachers predetermine what students need to know for tests and course grades. In such cases students never see the need or the "usefulness" of knowing. This encourages successful students to "pretend" that they have learned, since most can only recite words and do meaningless problems to illustrate their knowing.

Continuous Learning and STS: Specific Content versus Context

One of the exciting results of STS instruction is that learning is the major outcome—and learning is seen as a useful commodity that welcomes problems, dealing with them, and taking corrective actions. Unlike other reforms, STS involves all the steps of learning. Too many reforms focus almost entirely upon new curriculum structures, new and improved materials, and methods that teachers and students are expected to follow. For STS to work it requires open entry—something that is far more powerful than open-ended activities. This means that students must question, must be engaged with their own minds, and must experience a need to know.

Furthermore, real problems often cannot be solved; they are too complex, too difficult, too time-consuming, requiring information and skills not yet known. Perhaps a better term is "problem resolution." This requires teachers and students to limit the problem to smaller parts and to spend time working on them with the expectation that some resolution will occur. With a focus on problem resolution there is less emphasis on getting the right answer and committing it to memory for examinations. Psychologically there is the expectation that work will continue well beyond a particular instructional setting and involvement in some specific classroom or laboratory. Such a focus makes STS like the real world, where decisions must be made as to which problems to pursue, how the pursuit is organized and carried out, and what actions will be taken as a result of work on the problem. This feature of STS is one of its unique features; acquisition of concepts and skills occurs for a purpose—not merely because assignments are given, discussions are held, and examinations are administered where attention and recall ability are rewarded.

With experience throughout an STS course in using questions as a basis for study, students become skilled in following their interests and being challenged and involved with the interests of their peers. Similarly, they learn how to locate information to enhance their understanding and skills necessary for conducting experiments to test their ideas about possible problem resolutions. Open-entry allows students to formulate real questions, which in turn assist in generating and maintaining their motivation both inside and outside the classroom.

When students are in control of their learning instead of it being determined by the instructor as the "definer" of course content, they are empowered to think, to propose responses, to devise tests, to amass evidence, to communicate their ideas, report experimental results, provide and share the evidence for their interpretations, and use it to take actions to resolve problems. Such actions and abilities suggest that they will be useful outside the formal education session and beyond formal schooling. Students naturally become life-long learners because of such practice with using their own interests and knowledge base to deal with problems they choose.

In many situations STS nevertheless is defined as a specific content to be mastered by students. Unfortunately, this more limited view will rob most of the opportunity to learn in new and more useful ways. Just as courses can limit learning, so can thinking of

STS as a special field of study or inquiry. It is tempting in such situations to define the field in terms of a series of courses—all organized with textbooks, activities, and typical assessments based on remembering what was taught. This may especially be the case when STS is interpreted as science and technology studies, with their elaborate methodologies and scholastic terminologies.

When STS is used as a vehicle for teaching, students learn because the situation (context) engages their minds and invites involvement outside the classroom or laboratory. This corrects the problem of students not really "learning," but rather just "playing the game" needed to do well in the eyes of an instructor. Remembering details for tests is certainly not preparation for life-long learning, unless we assume that there will be an instructor dictating what is important to learn and requiring students to illustrate their learning by repeating and using their memories.

The advantages of STS certainly include the observation that such instruction assists in producing students who are thinkers, analyzers, designers of experiments, and communicators with others about their results, in contrast to students who are expected only to parrot the skills and concepts presented or heard. No reform will be possible, until we discontinue this view of instruction and related assessments of learning. If life-long contextual learning as typically practiced in STS classrooms is achieved, then its continued development throughout a lifetime is highly likely. This is what we desire for all the students we teach; STS makes it easier to succeed.

STS and the Constructive Perspective

Ernst von Glasersfeld, a prominent educator, defines constructivism as a way of knowing (1988). It is a theory emphasizing that meaning is not transmitted from one human to another. Instead meaning is developed individually in the mind of each person. Prior knowledge (including conceptions of the natural world not accepted by scientists) is often accepted without analysis or evidence, and simple direct teaching finds it difficult both to alter received knowledge or impart new meaning. In contrast, STS uses a constructivist perspective for learning, knowing, and the construction of meaning.

Contemporary science education research focuses more on students than teachers. With the emphasis on the learner, we see that learning is an active process occurring within and influenced by the learners as much as by an instructor, the curriculum, or the school. From this perspective, learning outcomes do not depend solely on what the teacher presents. Rather, they are an interactive result of what information is encountered and how the student processes it based on perceived notions and existing personal knowledge. All learning is dependent upon language and communication. Table 6.4 illustrates the role of communication in constructivist classrooms in different learning situations.

Von Glasersfeld (1988), as a prominent constructivist, claims that the existence of objective knowledge and the possibility of communicating that knowledge by means of language have traditionally been taken for granted by educators. During the last three decades faith in objective scientific knowledge has served as the unquestioned basis for most of the teaching in K-12 schools as well as in institutions of higher education.

Table 6.4
A Constructivist/STS Grid for Illustrating Instructional Strategies

Who	*Problems*	*Responses*	*Results*
A. Individual student	Identifying problem	Suggesting response	Self-analysis
B. Pairs of students	1. Comparison of ideas 2. Resulting questions	Agreeing on approach to problem(s)	Two-person agreement
C. Small group review	1. Consider different interpretations 2. Achieve consensus	1. Consider different responses 2. Achieve consensus	Small group consensus
D. Whole class (local community)	1. Discussion 2. Identify varying views	Acting to gain consensus	Whole class agreement
E. Science community	Comparison of class view vs. those of scientists	Comparison of class views vs. those of scientists	Consensus/new problems/ actions

Von Glasersfeld argues, however, that instead of presupposing knowledge has to be a representation of what exists, we should think of knowledge as a mapping of what turns out to be feasible, given human experience. If we were to do so, he suggests that profound changes in the way we teach our children would result and that students would study for real learning and understanding rather than for recalling what teachers directed. In the constructivist perspective students would be more central to the process; curriculum materials are designed more effectively; and teachers realize that rote learning and repeated practice are not likely to generate real understanding and useful knowledge.

Constructivist teachers understand that knowledge cannot simply be transformed by means of words. Explaining a problem will not lead to understanding unless the learning has an internal scheme that maps onto what a person is hearing. Knowledge is not acquired passively, because learning is a product of self-organization and reorganization. Constructivist teachers of science thus promote group learning, where two or three students discuss approaches to a given problem with little or no interference from a teacher. What happens to and with such small groups of students can be used as a whole class arrives at consensus about various small group analyses.

Insofar as learning and knowledge are instrumental in establishing and maintaining a student's equilibrium, they are adaptive. Once this way of thinking takes hold, teachers change their view of problems and solutions. It is no longer possible to cling to the notion that a given problem has only one solution. It is also difficult to justify conceptions of right and wrong answers. STS constructivist teachers would rather explore how students see the problem and why their paths toward solutions seem promising to them.

Constructivism has major implications for curriculum frameworks and textbooks. Most existing courses and textbooks are organized around concepts that someone else has decided students should master. They are not designed to encourage an open-ended investigation of topics that interest students. In a constructivist classroom the function of textbooks and other support materials are very different from most current practices. Instead of defining what should be learned, to what depth, and in what order, textbooks ought to provide a source of data that students would be expected to prove rather than ingest. Ideally, such books might include a diversity of perspectives, even conflicting opinions. They should not strive to complete but rather leave to students the task of deciding how and where to search for additional data. Students should also confront the task of evaluating sources; for this reason textbooks ought to contain errors and information of dubious quality. Fortunately, most textbooks contain ample amounts of dubious information, even without the benefit of constructivism.

Textbooks should ask questions, pose problems, suggest areas for inquiry and actions to advance inquiry. Textbooks should help students—just as teachers should, but this help should not involve simply telling the reader what to believe or what facts to memorize for an exam. They should ask for student perceptions, offer ways of verifying them, encourage interactions with others, provide extension activities, and suggest ties to local conditions. Students should learn that textbooks can enhance learning and not represent the parameters of information defined for a given course by a teacher.

The Wave of the Future

The biggest paradigm shift required by constructivism concerns new roles for teachers and new demands placed on teacher education. Teachers must learn to master a broader set of roles, behaviors, and strategies. The best teachers would see themselves as active learners; they would always strive to understand better their own teaching and the effect it has on student learning. This is a continuing process—a type of learning that is never complete and never dull. Constructivist teaching always provides an exciting adventure; courses where traditional teaching is perceived as routine or unrewarding can be transformed into stimulating challenges (Lochhead, 1991). Such teaching is essential when STS is viewed as a reform in education.

Most science teachers like the science they learned in school and proceeded to college majoring in science because of this liking and the success they experienced with their subject. Science teaching in secondary school is modeled after what teachers experienced during their own college preparation. Most college faculty rarely employ constructivist instructional practices. Professors "profess"; they give lectures on topics they view as important. They expect students "to learn" the material presented in lecture and assess student learning largely on the basis of factual recall. Introductory laboratories emphasize following instructions and getting results, where the desired result is known well before the "investigation" begins. While university-based scientists often blame education programs for what they believe are poorly trained teachers, most prospective science teachers take far more courses in science than in education. Breaking out of the old mold will require either a change in the manner in which university science is taught or a reduction in the degree to which prospective teachers stress science content courses at the university.

STS provides a vision for change. It also provides an example of how the new National Science Education Standards (1996) can be implemented. STS instruction is successful in changing teacher and student perceptions, which also affect student learning in a variety of domains, including concept mastery, process skills, use of concepts and processes in new contexts, the nature and history of science, creativity, and developing and maintaining positive attitudes. STS provides a mechanism to accomplish these reforms that are at the top of our education agenda. This mechanism is related more to instruction and teaching than it is to curriculum structures, which means it also requires new ways of assessing real learning. Fortunately, more and more materials are being produced at every level that illustrate desired instructional strategies as well as curriculum frameworks to illustrate the current educational reforms generally advocated, while moves to more authentic ways of assessing learning are also developing swiftly, especially when STS reforms are implemented successfully.

References

College Board. 1990. *Academic Preparation in Science*, 2nd ed. New York: College Board.

Dewey, John. 1925. *Experience and Nature*. Chicago: Open Court.

Harms, Norris C. 1977. *Project Synthesis: An Interpretive Consolidation of Research Identifying Needs in Natural Science Education*. (A proposal prepared for the National Science Foundation.) Boulder, CO: University of Colorado.

Iowa Assessment Handbook. 1998. Iowa City: University of Iowa, Science Education Center.

Kellerman, Lawrence R., and Liu, Chin-tang. 1996. "Enhancing Student and Teacher Understanding of the Nature of Science via STS." Pp. 139–48 in Yager, ed., 1996.

Kowal, Jerry. 1991. "Science Technology and Human Values: A Curricular Approach." *Theory into Practice*, vol. 30, no. 4 (autumn), pp. 267–72.

Liu, Chin-tang, and Yager, Robert E. 1996. "An STS Approach Accomplishes Greater Career Awareness." Pp. 149–62 in Yager, ed., 1996.

Lochhead, Jack. 1991. "Making Math Mean." Pp. 75–87 in Ernst von Glasersfeld, ed., *Radical Constructivism in Mathematics Education*. The Netherlands: Kluwer.

Lochhead, Jack, and Yager, Robert E. 1996. "Is Science Sinking in a Sea of Knowledge? A Theory of Conceptual Drift." Pp. 25–38 in Yager, ed., 1996.

Lux, Donald G. 1984. "Science and Technology: A New Alliance." *Journal of Epsilon Pi Tau*, vol. 10, no. 1 (spring), pp. 16–21.

McComas, William F. 1996. "The Affective Domain and STS Instruction." Pp. 70–83 in Yager, ed., 1996.

McShane, Joan B., and Yager, Robert E. 1996. "Advantages of STS for Minority Students." Pp. 131–38 in Yager, ed., 1996.

National Science Teachers Association. March 1990. *Qualities of a Scientifically and Technologically Literate Person*. Task Force Report. National Science Teachers Association, 1990–1991.

Olson, Eric, and Iskandar, Srini. 1996. "Enhancement of Opportunities for Low-Ability Students with STS." Pp. 109–18 in Yager, ed., 1996.

Penick, John E. 1996. "Creativity and the Value of Questions in STS." Pp. 84–94 in Yager, ed., 1996.

Perry, William G. 1970. *Forms of Intellectual and Ethical Development in the College Years: A Scheme*. New York: Holt Rinehart and Winston.

"Science/Technology/Society: A New Effort for Providing Appropriate Science for All (Position Statement)." Pp. 47–48 in *NSTA Handbook*, Washington, DC: National Science Teachers Association.

Snow, Charles P. 1959. *The Two Cultures and the Scientific Revolution*. New York: Cambridge University Press.

U.S. Department of Education, National Center for Statistics. 1991. *The Condition of Education 1991*, vol. 1: *Elementary and Secondary Education*. Washington, DC: U.S. Department of Education, National Center for Statistics, pp. 38–39.

Varrella, Gary F. 1996. "Using What Has Been Learned: The Application Domain in an STS-Constructivist Setting." Pp. 95–108 in Yager, ed., 1996.

von Glasersfeld, Ernst. 1988. *Cognition, Construction of Knowledge, and Teaching*. Washington, DC: National Science Foundation.

Whitehead, Alfred North. 1925. *Science and the Modern World*. New York: Macmillan.

Yager, Robert E., ed. 1993. *What Research Says to the Science Teacher*. Vol. 7, *The Science, Technology, Society Movement*. Washington, DC: National Science Teachers Association.

————, ed. 1996. *Science / Technology / Society as Reform in Science Education*. Albany, NY: SUNY Press.

Yager, Robert E., Myers, Lawrence H., Blunck, Susan M., and McComas, William F. 1990. "The Iowa Chautauqua Program: What Assessment Results Indicate about STS Instruction." Pp. 133–47 in Dennis W. Cheek, ed., *Technology Literacy V: Proceeding of the Fifth National Technology Literacy Conference, Arlington, VA; February 2–4, 1990*. Columbus, OH: ERIC Clearinghouse for Science, Mathematics, and Environment Education.

Ziman, John. 1980. *Teaching and Learning about Science and Society*. Cambridge: Cambridge University Press.

Zuga, Karen F. 1996. "STS Promotes the Rejoining of Technology and Science." Pp. 227–37 in Yager, ed., 1996.

7

STS from a Policy Perspective

ALBERT H. TEICH

Politics is often the societal arena in which science- and technology-related issues and problems are worked out. Whether at the local, state, federal, or international levels, science and technology are vitally implicated by virtue of their centrality to economic, environmental, legal, and a host of other issues. Government, especially at the national level, is also a financial supporter of scientific research and technological development. All this and more is an object of study by political scientists interested in science and technology. While the policy dimensions of STS are important objects of academic study, for Albert Teich, it is "the very essence of what I and my colleagues do."

Teich is head of the Directorate for Science and Policy Programs of the American Association for the Advancement of Science (AAAS). Now over 150 years old, the AAAS is the largest professional scientific organization in the world. In addition to furthering the work of scientists—which includes the publication of the weekly magazine *Science*—it has among its objectives fostering scientific freedom and responsibility, increasing public understanding of science, and providing scientific and technical knowledge to decision-makers, including decision-makers in the U.S. Congress. Thus, as Teich puts it, the AAAS and his program in particular are at the interface of science, technology, and society. In addition to serving as an AAAS administrator of policy programs, Teich is a scholar in his own right, having published numerous articles and books in the STS field. Most notable is his edited collection of essays titled *Technology and the Future*, now in its 8th edition (2000).

Drawing on his many years of experience and powers as an astute observer of the STS field, Teich offers several illustrations of how STS perspectives inform his work. In particular, he focuses on the relationship between science and the law, accountability in scientific research, and the tracking of research funding. In these and other areas, he claims that STS knowledge helps to improve science and technology policy by enriching our understanding of those activities. To what extent does this claim appear well founded? Are there similar examples from other areas that could be added to Teich's list?

As science and technology become more and more important to the fabric of modern society, the role of government in mediating the relationships among science, technology, and society (STS) inevitably grows stronger. Among the ways in which this phenomenon is manifested is the increasing importance of scientific and technical information and expertise in decision-making in all of the branches and levels of government. It can also be seen in the growing government involvement in allocating resources, setting directions, and shaping the environment for research.

The need for a strong government role in science and technology policy seems clear, both because of the dependence of research (at least basic research) on public funding and because of the impacts of science and technology on the economy, the environment, and so many other aspects of society. Nevertheless, governments are often not well equipped to make decisions involving science and technology. And bad decisions can contribute to the erosion of democratic values as well as stunting or misdirecting the growth of science and technology.

Studying the relations between science and government is an important preoccupation of many STS researchers, but as an administrator in a large scientific organization based in Washington, D.C., I see the policy dimensions of STS not just as objects of study. STS—at least its policy side—is the very essence of what I and my colleagues do. The American Association for the Advancement of Science (AAAS), where I head the Directorate for Science and Policy Programs, exists at the interface of science, technology, and society.

Founded in 1848, AAAS is an organization of more than 140,000 individual members. Among its objectives are: furthering the work of scientists and facilitating cooperation among them, fostering scientific freedom and responsibility, improving the effectiveness of science in the promotion of human welfare, and increasing public understanding of science. The programs for which I am responsible serve these objectives in many different ways ranging from promoting dialogue between science and religion to improving the relations between law and science, to tracking research funding in the federal budget, to helping provide better technical information and expertise to decision-makers in the U.S. Congress and other parts of the federal and state governments.

What is life like on this interface and how can STS perspectives inform the activities of me, my colleagues, and others who operate in this environment? A few illustrations, drawn from among the many issues in which AAAS and the other organizations that play in the Washington science policy game are involved, may be the most useful way to answer these questions.

Science and the Law:
The Nature of Scientific Evidence

An increasing share of important cases that come before the courts in the United States involve scientific and technical considerations. These range from DNA "fingerprinting" technology in criminal cases (the O.J. Simpson case being the best known example) to the use of epidemiological and laboratory studies as evidence in environmental regulation and in civil lawsuits (such as those involving silicone breast implants). Most often, the opposing sides in such cases present evidence and interpretations that appear to be diametrically opposed. And the result, in many cases, is a decision that appears to be inconsistent with the widely shared scientific consensus.

Controversy in the legal and policy community has centered on what constitutes an appropriate standard for the admissibility of scientific evidence. For many years, the courts followed a rule based on a 1923 federal court decision that allowed scientific testimony into evidence if it was based on techniques or approaches that were "generally accepted" in the scientific community. Most judges regarded publication in a peer-reviewed journal as an indication of general acceptance. In 1975 Congress enacted a new set of federal rules of evidence broadening the standard to allow testimony based on ideas that were not widely embraced. Many state courts continued to follow the original rule, leading to much confusion and inconsistency among decisions.

My colleagues in the AAAS Program in Science and Law became involved in this controversy in late 1992 when a case providing an opportunity to clarify this situation reached the U.S. Supreme Court (Daubert et al. vs. Merrell Dow Pharmaceuticals, Inc.). Together with the National Academy of Sciences, AAAS filed an amicus curiae brief urging the court to "uphold the broad authority of trial judges to exclude putatively scientific evidence that does not, according to the standards applied by the scientific community, have the earmarks of scientific reliability." The brief asserted that judges should rely on well established criteria to evaluate the validity of scientific evidence and to screen it for admissibility. Such criteria would include, but not be limited to, peer review (American Association for the Advancement of Science and National Academy of Sciences, 1993).

Several other amicus briefs were also submitted. One was prepared by an independent group of STS scholars and scientists. This brief argued that, while peer review has many virtues, it is neither a "guarantor of scientific certainty," nor an appropriate gatekeeper for expert witnesses in litigation (Chubin, Hackett et al., 1992). Another stemmed from the work of the Carnegie Commission on Science, Technology, and Government, also arguing against the notion that peer review should be the sole condition for admissibility of scientific evidence (Carnegie Commission, 1992). While these briefs were somewhat different in outlook, they all drew on the STS literature to discuss the nature of scientific knowledge as well as the virtues and limitations of various means of establishing the validity and reliability of evidence.

The Supreme Court decision, handed down in June 1993, took account of these STS perspectives. It threw out the "general acceptance" criterion in favor of a standard that gives judges more latitude, requiring them to assess not the conclusions but "whether the reasoning or methodology underlying the testimony is scientifically valid . . . and can be applied to the facts in issue." Thus controversial findings cannot be excluded just because they have not been peer reviewed, but at the same time, judges will have the latitude to throw out so-called junk science, that seems to carry the mantle of science but does not derive from scientific methodology (Mervis, 1993). Overall, the Daubert decision and a subsequent decision expanding upon it (Joiner vs. General Electric Co.), place the burden of deciding the reliability of scientific testimony on judges, calling on them, in effect, to use much of what STS has learned about the construction of scientific knowledge.

GPRA and Accountability in Scientific Research

Among the trends shaping government policy in recent years has been an increasing focus on accountability—assuring that the public gets its money's worth from government programs. One expression of this focus is the Government Performance and Results Act, GPRA, passed by Congress and signed by President Clinton in August 1993. GPRA is intended to shift the focus of government agencies and officials from inputs (e.g., how much money is being spent) to outputs. (What kind of results are being achieved? How well are programs meeting their objectives?) The legislation requires federal agencies to set long-term goals as well as annual targets and to report each year on how their actual performance compares to their targets (also known as performance goals).

While the ultimate purpose of this legislation is a worthy one, its application to the federal government's R&D programs, particularly its basic research, is a source of concern to many scientists and people involved in science policy. How government agencies might measure the performance of programs that support basic research is not a straightforward matter. How does one define outcome goals for an activity that is inherently long-term and uncertain? And how does one judge success or failure in meeting those goals, particularly on an annual basis?

Some scientists have expressed concern that GPRA might drive federal agencies away from support of basic research programs and toward applied research whose outcomes and performance goals can be specified more clearly and in which progress can be more easily measured. While this is certainly not the intent of the legislation (which applies to the whole government, not just research agencies), it is just as certainly not an idle concern. Well intentioned laws have a way of producing unintended consequences, some of which may be in conflict with the objectives of those who framed the laws. At the same time, some in the science policy field have argued that GPRA will force research funding agencies and researchers to do things they should have been doing on their own in recent years—that is, communicate the value of research more effectively with policy-makers and the public.

Because GPRA's focus is on government agencies, AAAS has not been a central player in the activities surrounding its implementation. The key actors in GPRA implementation so far have been program managers and administrators in federal agencies. My directorate at AAAS has, however, provided an important forum for scientists and policy-makers to discuss concerns about the law's application to science agencies, for example, at a symposium on the subject at the 1994 AAAS Colloquium on Science and Technology Policy (Teich, Nelson, and McEnaney, 1994) and, more recently, at a seminar for congressional staff in the fall of 1998.

In the research oriented agencies, such as the National Science Foundation (NSF), the products of STS define the background against which this implementation is taking place. Evaluation of research programs may involve peer review, bibliometrics, and a variety of other methods shaped and studied by STS researchers. Whether they are explicitly acknowledged or not, the models of innovation (linear, chain-linked) and the understanding of how science progresses developed (or in some cases debunked) by STS researchers underlie the basic assumptions on which goals will be formulated and against which performance will be measured. It is unlikely that these methodologies, understandings, and models will be applied appropriately in all, or even most, instances as GPRA evolves. AAAS and other organizations in which STS researchers and science policy-makers participate will continue to play a key role therefore in connecting these two worlds.

Choosing among Disciplines

One of the enduring and best-known elements of the AAAS work in science policy over the past two decades is a program that tracks research funding in the federal budget. My own career with AAAS began in this activity and I continue to play a significant role in it. We seek to provide timely, accurate, and objective information about research funding to researchers, administrators, government officials, members of Congress, and staff, journalists, and many others. We play the role of informed observers and stop short of making recommendations. The R&D Budget and Policy Program gives AAAS an excellent vantage point from which to watch the ebb and flow of science policy through the prism of budgetary politics.

One of the continuing themes in this arena is the issue of how to set priorities for research. Often it is posed in terms of the need to choose among disciplines: deciding how much federal funding should go to molecular biology, how much to astrophysics, how much to social psychology, and so on. Decisions on funding at the macro level (but not the micro level of individual grants) are necessarily made in a political environment. Funds are allocated to the various federal agencies that support research through a process that can seldom be described as rational and the results are often questioned by scientists and politicians alike (Teich, 1994).

Physical scientists look enviously at the increases granted to the National Institutes of Health (NIH) year after year and complain that these increases are coming at the expense of physics, chemistry, and geology. Biologists look at the price tags on the tools of "big science" fields like high energy physics—for example, huge accelerators for subnuclear particles—and question the cost-benefit ratios. And political leaders express dismay at the fact that the scientific community is fragmented and fails to provide useful guidance on its priorities, leaving the tough problem of allocating scarce resources to them.

Scientists have, over the years, attempted to develop schemes for the scientific community to recommend research priorities across disciplines. In the 1960s, Alvin Weinberg, then director of Oak Ridge National Laboratory and a member of the president's Science Advisory Committee, proposed "criteria for scientific choice" (Weinberg, 1963). More recently, a committee of the National Academy of Sciences chaired by its former president, Frank Press, prepared a report on *Allocating Federal Funds for Science and Technology*, suggesting a different set of criteria. None of these has gained acceptance in the budget process (National Research Council, 1995).

The STS community has done relatively little with the issue of priority-setting directly. Neither the term nor any of its common synonyms (allocation, choice) appears in the index of the latest *Handbook of Science and Technology Studies* (Jasanoff et al., 1995). Nevertheless, STS has contributed to understanding the context of priority-setting for research by enriching the picture of science as a fundamentally political activity. As Cozzens and Woodhouse (1995) describe in their chapter in that *Handbook*, STS research has examined the ways in which the distribution of interests in society affects the generation and utilization of scientific knowledge. The notion that priorities for science can somehow be set by the scientific community independently of its political setting is belied by this research. This reality is something that I and my colleagues in the AAAS R&D Budget and Policy Program see every day.

STS and the Future of Science Policy

These are but a few of the many possible examples I might have chosen to illustrate the ways in which my work and that of others at AAAS draws upon the ideas and approaches of STS in helping to shape science policy. In a broad sense, STS provides a framework in which we can think about science and technology, and their interactions with society. But science policy faces many challenges from a dramatically altered political and economic situation in the world and a range of socio-technical problems, such as global climate change, that have few precedents in human history. The practitioners of science policy in government, in organizations like AAAS, and in a host of other institutions will need a new generation of creative thought if we are to rise to these challenges.

The academic study of STS can contribute to improving science policy by enriching our understanding of its subject matter. This does not imply that all or even most STS research should aspire to be directly relevant to science policy. Like the scientific fields that are the objects of its study, STS itself is socially constructed. It develops through its own internal dynamics under the influence of its social environment. STS creates a part of the background for the stage on which science policy is played out. Its practitioners interact with those who shape policy and from time to time become policy-makers themselves. Its students— those who major in it, of course, but even more importantly the much larger numbers of those who take STS courses as part of a science, engineering, or liberal arts curriculum—bring to their roles in society an enhanced appreciation of the fact that science and technology are not something set apart from society, but rather are integral elements of it. Gradually, the concepts of STS become part of the policy discourse. To the extent that STS researchers recognize the importance of their field to policy-making and make efforts to communicate their ideas more clearly to nonspecialists than has often been the case, the influence of STS scholarship is bound to grow.

References

American Association for the Advancement of Science and the National Academy of Sciences. 1993. Brief as Amici Curiae in support of respondent, in the Supreme Court of the United States, October Term, 1992, Daubert vs. Merrell Dow Pharmaceuticals, Inc. No. 92–102.

Carnegie Commission on Science, Technology, and Government. 1992. Brief as Amicus Curiae in support of neither party in the Supreme Court of the United States, October Term, 1992, Daubert vs. Merrell Dow Pharmaceuticals, Inc. No. 92–102.

Chubin, Daryl E., Hackett, Edward J. et al. 1992. Brief as Amici Curiae in support of petitioners, in the Supreme Court of the United States, October Term, 1992, Daubert vs. Merrell Dow Pharmaceuticals, Inc. No. 92–102.

Cozzens, Susan E., and Woodhouse, Edward J. 1995. "Science, Government, and the Politics of Knowledge." Pp. 533–53 in Jasanoff et al., eds., 1995.

Jasanoff, Sheila, Markle, Gerald E., Petersen, James C., and Pinch, Trevor, eds. 1995. *Handbook of Science and Technology Studies*. Thousand Oaks, CA: Sage.

Mervis, Jeffrey. 1993. "Supreme Court to Judges: Start Thinking Like Scientists." *Science*, vol. 261 (2 July), p. 22.

National Research Council. 1995. *Allocating Federal Funds for Science and Technology*. Report of the Committee on Criteria for Federal Support of Research and Development. Washington, DC: National Academy Press.

Teich, Albert H. 1994. "Priority-setting and economic payoffs in basic research: An American perspective." *Higher Education*, vol. 28, no. 1 (July), pp. 95–107.

Teich, Albert H., Nelson, Stephen D., and McEnaney, Celia. 1994. Part 6, "Perfomance Assessment for Research and Development Programs." Pp. 253–305 in *AAAS Science and Technology Yearbook*. Washington, DC: American Association for the Advancement of Science.

Weinberg, Alvin M. 1963. "Criteria for Scientific Choice." *Minerva*, vol. 1, no. 2 (winter), pp. 159–71.

III

Critiques

8

STS on Other Planets

RICHARD E. SCLOVE

In a democratic society the STS perspective has the potential to contribute in myriad ways to better science and technology decision-making, as has already been suggested by several of the authors in this collection. Recently, however, much of the scholarly STS literature has focused on the "social construction" of scientific knowledge and technological change, sometimes to the exclusion of concern for the "social consequences" of science and technology. Richard Sclove is particularly concerned that STS not lose sight of its sociopolitical roots. He thus argues for "socially engaged STS research," by which he means that STS professional institutions, teachers, and even students should work to make science and technology responsive to democratically decided social concerns.

Sclove's plea is more than idle academic rhetoric, since he is the founder and research director of the nonprofit Loka Institute, which has as its mission the promotion of a democratic politics of science and technology decision-making that includes a strong grassroots, worker, and public-interest voice. To that end he is also the author of *Democracy and Technology* (1995), which not only argues for this position, but also offers "design criteria for democratic technologies."

In the essay that follows, Sclove "fantasizes" about an imaginary planet where there is active democratic participation in the processes of scientific and technological decision-making. Shifting out of this rhetorical reverie, he notes that this image is, in fact, a very real one, even if limited in scope, and that it can be found in the realization of Dutch "science shops" and Danish "consensus conferences." Although differing in approach, in both cases the intent is to provide expanded knowledge and to allow greater participation for the general public with regard to scientific and technological issues.

In concluding his essay, Sclove invites professors and students alike to become more actively engaged in socially responsive programs and institutions. In reading this essay, we may want to compare Sclove's approach to the question of science and technology policy- and decision-making with those of Lars Fuglsang and Albert Teich earlier in this collection. It would also be a useful exercise to identify those institutions and opportunities available locally where such active participation is possible or, barring such opportunities, to consider how to create just such possibilities.

Knowledge from Nonexperts

The practice of science and engineering occurs preeminently in laboratories, so it is fitting that those of us concerned with the social aspects of science and technology preoccupy ourselves with those haunts. Yet periodically, when suffering snow blindness from staring at men in dazzling white coats, I close my eyes and fantasize about another planet: Planet XI. Like Earth, Planet XI is a place in which knowledge and technical artifacts are socially contrived, but in other respects it is utterly bizarre and exotic.

For example, on Planet XI a startling amount of knowledge about how the world works is produced by social groups comprising nonexperts—that is, ordinary women and men. Sometimes they are organized according to their occupations (a little bit like our trade unions), sometimes according to their social concerns (like our environmental or women's groups), and sometimes according to where they live (like our community and grassroots organizations).

Some of these groups produce knowledge entirely by themselves. For instance, if they fear that they have been poisoned by polluted water, they conduct their own surveys and empirical examinations to find out whether, how, and why. Farfetched as this sounds, they are able to do this without the benefit of university educations, research grants, or laboratory facilities.

But in other cases they produce knowledge in close collaboration with professionally trained researchers. Yes, it is hard to believe that there could be a place where men and women with professional credentials would even talk, much less cooperate actively, with others less educated. But on Planet XI, I insist, it is so.

For instance, in one nation on Planet XI, most universities have established a set of research centers whose sole purpose is to facilitate studies conducted with or for popular organizations. Thus, on Planet XI, understanding how knowledge is socially produced sometimes entails studying laboratories, but it also means spending time with all kinds of women and men in all kinds of social settings. On Planet XI, knowledge creation knows no sharp geographic, class, or other social boundaries.

Even on Earth, science and technology are not, of course, self-contained or insular enterprises; they are strongly influenced, for example, by government policies. But since the only kinds of people who significantly influence those policies are the same people who otherwise wear white coats and busy themselves with laboratory equipment, studying science policy-making on Earth hardly requires shifting one's gaze from the laboratory's customary denizens. Thus, it is both a relief from tedium—and yet also a bit shocking—that on Planet XI many other kinds of people influence science and technology policy-making.

For instance, there is another nation on Planet XI that, realizing that knowledge and know-how are not only socially constructed but also have profound social repercussions, convenes panels of laypeople—that's right, everyday folks from all walks of life, including school teachers, homemakers, and street sweepers—to publicly interrogate men and women in white coats and then reach their own policy conclusions. These lay panelists' judgments have influenced popular political deliberations, business decisions, and government policies.

You might well imagine that this process is not only costly but leads to ludicrously ill-informed judgments. But a broad cross section of the nation's members—including its political and business leaders—claim that these irrational participatory methods actually result in greater social justice and even in real economies. This occurs, according to them, because there is relatively little costly opposition to innovation, insofar as a wide range of social concerns are reasonably well reflected in prior research and development (R&D) and government policy decisions.

In several nations on Planet XI, programs have begun to be established through which workers and consumers can even participate directly in designing alternative technologies better adapted to their life circumstances and aspirations. Workers, for instance, have consistently demonstrated both an interest and impressive capabilities in helping to devise production technologies that are not only efficient but also maintain safe, high wage/high skill jobs, protect the environment, and result in high quality products or services.

Many university students on Planet XI pursue educations and careers no different from the conventional student trajectories familiar on planet Earth, but others choose to become actively engaged in the preceding participatory activities as an integral aspect of their studies. For instance, one Planet XI university has a community research center located *within* its academic technology and society program. The center is staffed by professors of science, technology and society (STS), who also teach courses on participatory research and on participatory approaches to technological design.[1]

Students who take these courses receive credit for conducting participatory community research projects. Their projects, in turn, influence the university to adopt new courses that reflect community concerns (such as sustainable economic development) and to establish new, socially oriented, interdisciplinary research programs that include faculty from many different departments and programs throughout the university. This university's STS professors themselves hold graduate degrees in either natural science, engineering, or social science. Disciplinary credentials turn out to be of secondary importance, however, because over time all the professors have become generally familiar with one another's disciplines.

Expert and Nonexpert Criticism

To read the mainstream STS literature currently being produced back home on Earth, one would have to conclude that Planet XI exists only in my fevered imagination. But actually, Planet XI is a real place. In fact, it is the third planet out from the center of our own solar system. For instance, in 1995 I made a brief tour of two of the nations on Planet XI; they are named "Denmark" and "the Netherlands."

During this trip I was privileged to deliver a plenary address to the national meeting of the Dutch "science shops." The meeting was attended by staff from the Netherlands' thirty-eight university-

based community research centers, which together produce more
than a thousand studies each year in response to requests from
community groups, trade unions, public-interest organizations, and
local governments.[2]

Other science shops, or related community research centers
(not always based in universities), now exist in many other nations,
including Denmark, Austria, Germany, England, Ireland, Norway,
the Czech Republic, Canada, and the U.S., although the Dutch system is the oldest and mostly highly evolved. In the developing world
there is a somewhat analogous international network of indigenous
knowledge resource centers; its newsletter is published in The
Hague.[3]

I also met with staff from Teknologi-Rädet (the Danish Board
of Technology), who since 1987 have conducted about twenty "consensus conferences" in which panels of everyday citizens become
intensively informed on selected topics in science and technology
policy and then, after participating in a public forum, announce
their judgments at national press conferences that are attended by
members of Parliament (Sclove, 1996; Joss and Durant, 1995).[4]

I spent a day with several professors at Aarhus University,
who are among the world's leading practitioners in designing new
technologies collaboratively with workers (*Computers in Context*,
1995). I was hosted for another day at the Danish Technological
University in Lyngby, where indeed there is a thirteen-year-old science shop located within an STS program and staffed by Professors
Michael Søgaard Jørgensen and Børge Lorentzen.

One comes naturally to the question of why these, as well as
other real-life examples that seemingly represent an important
thrust toward democratizing science and technology, are so little
considered within the conventional STS literature. The first Danish
consensus conference was held in 1987, but the leading academic
STS journals—such as *Science, Technology & Human Values* and
Social Studies of Science—have not discussed these procedures.[5]
How do the reports produced by Danish lay panels compare sub-
stantively with those produced by conventional technocratic
approaches to technology assessment? Is their social and political
impact typically greater or less? The bulk of the STS community
has apparently not found such questions of interest.

During the mid-1980s Professor Loet Leydesdorff and colleagues published several illuminating studies of the main science
shop at the University of Amsterdam (Leydesdorff and Van den
Besselaar, 1987; Zaal and Leydesdorff, 1987). But at the time there

were already about a dozen other science shops scattered throughout the Netherlands. What of them? Indeed, since that time the
number of Dutch science shops has tripled, but apparently no one
in the STS community has found this vibrant effort to democratize
university research capabilities worthy of serious attention. In fact,
when the very shop that Leydesdorff and his colleagues studied was
recently shut down owing to university budget constraints, did a
single person from the STS community know, care, or do anything
to try to help?

How do the several dozen remaining Dutch science shops vary
from one another? How are participating students' career decisions
affected? Do the shops appreciably influence faculty research programs? What is the social impact of the shops' research? How does
their social utility and cost efficacy compare with that of conventional research systems? How do science shops in various countries
reflect the different circumstances of their origin? Could science
shops and the popular constituencies they serve evolve into a grassroots foundation for challenging other, nondemocratic science and
technology institutions? Is the Internet permitting transnational
collaborations among science shops to emerge?

No one knows the answer to these and a hundred other such
questions, for the simple reason that no one has asked them. The
answers would not merely be of academic interest; they could help
provide a basis for maintaining and greatly extending the practice
of community-based research (Sclove et al., 1998, esp. chap. 3, finding 18).

The pioneering anthologies on participatory research have all
been published by Third World activists or by social change-ori-
ented sociologists, not by members of the STS community (Fals-
Borda and Rahman, 1991; Park et al., 1993; Nyden et al., 1997).
Likewise, the pioneering anthologies on participatory design in the
workplace were compiled by Computer Professionals for Social
Responsibility—an activist group, not an STS organization (e.g.,
Schuler and Namioka, 1993; Chatfield et al., 1998). The latter
anthologies are extremely useful, but other questions remain to be
asked. For instance, if workers and users should participate in
technological design, what about affected *non*-users? What are the
cultural, institutional, and legal barriers to participatory design,
and what types of political strategies might be used to soften them?
(Sclove, 1995; Sandberg et al., 1992)

Every year in the United States a majority of new STS graduate students, and many undergraduates, arrive on campus moti-

vated primarily by awareness of some particular deep social problem involving science or technology. They want to study that problem, and to contribute constructively and actively toward addressing one or another real social ill. Do our current STS programs nurture that eminently worthy desire?

For the most part, no. These admirably motivated students are coopted into courses and research programs whose inadvertent (?) thrust is to remake their social commitment into a commitment to largely idle scholarship instead. This is good for academic careers, perhaps, but not for society. STS—as a codified profession, field, or discipline—is now near-perfectly accomplishing just what the late social theorist Michel Foucault claimed disciplines normally do: producing docilely functional bodies.

Similarly, the academic STS community's recent, intense preoccupation with establishing that technologies are contingent social *products* (a theoretical point that was actually pretty well established in the 1970s by social historians of technology and by appropriate technology practitioners) has meant that few in the STS community are studying the other half of the coin: the social *consequences* specific to particular technologies and technological complexes. The relative inattention to consequences has been noted, for instance, by sociologist Claude Fischer (1987), technological change theorist Everett Rogers (1983), and urban historian Christine Rosen (1989).

The embarrassing truth is that when I want to learn about the social consequences of emerging technologies, I do better canvassing human interest stories by *New York Times* reporters than reading anything in the leading STS journals. Recently in the U.S., the most influential scholarly claims about the social and political implications of technology have been made by Harvard University political scientist Robert Putnam (1996), who never cites any STS literature and has never published in STS journals.

A few others in the academic STS community have called attention to various expressions of depoliticization within our field—famously symbolized by the recent the shift in meaning of "STS" from "science, technology and society" to "science and technology studies"—but little has yet changed as a result of these critiques.[6]

So, why is STS relegating overt attention to democratizing science and technology to a back burner? One obvious hypothesis is that such attention would directly challenge current social power relations and so risk currying disfavor within the corridors of power, including those that provide funding. Servants of power are rewarded in our societies; challengers are frequently punished.

This hypothesis is unfashionably straightforward and simple, but there is also some evidence to support it. For example, two of the most gifted and inspiring STS professors with whom I studied as a beginning graduate student in the 1970s were David Noble and Langdon Winner. Both were politically engaged, and both were, not coincidentally, denied tenure by MIT. Did these spectacularly unjust and irrational decisions function as early warning shots across the bow, teaching other aspiring STS scholars the career risks they might run if they did not depoliticize their research and teaching programs?

Perhaps one way to start reversing this socially damaging, climatic chilling within our field would be for socially concerned STS professors—or, better yet, the leading STS professional societies—to establish standby mechanisms for quickly mobilizing external support to colleagues whose political commitments are jeopardizing their careers. We could also establish prizes to recognize and reward socially engaged research and teaching. Meanwhile, university students interested in becoming involved with socially engaged STS research, or in helping directly to make science and technology more responsive to democratically decided social and environmental concerns, can do so by applying to volunteer or intern with such nonprofit organizations as the Loka Institute, Council for Responsible Genetics, or Computer Professionals for Social Responsibility.[7]

Notes

1. As noted later in this essay, "STS" is a contested acronym. It originally meant "science, technology and society," but in some circles it now denotes "science and technology studies."

2. Sclove (1995, pp. B1–B3) (this essay is available at <http://www.Loka.org/alerts/loka.2.5a.txt>); or Sclove, Scammell, and Holland (1998, esp. sect. 2.13) (this report is available at <http://www.loka.org/crn/pubs/comreprt.htm> or it may be ordered directly from the Loka Institute, P.O. Box 355, Amherst, MA 01004 USA; tel. [413] 559–5860; e-mail <Loka@Loka.org>).

3. Directories of science shops and related community research centers worldwide are available via the Community Research Network Database at <http://www.loka.org/crn/index.htm> and via Delft Technical University at <http://www.bu.tudelft.nl/wetensch/lsw/ehome.htm#start>.

Information about indigenous knowledge resource centers is available from the Center for International Research and Advisory Networks (CIRAN), P.O. Box 29777, 2502 LT The Hague, The Netherlands, e-mail <ikdm@nuffic.nl.> or <http://www.nuffic.nl/ciran/ik.html>.

4. Sclove (1996) is also available at <http://www.loka.org/pubs/techrev.htm>; and Joss and Durant (1995). Information about the first consensus conference in the United States, organized in April 1997, is available at <http://www.loka.org/pages/panel.htm>.

5. Guston (1999) represents a recent, welcome exception.

6. See, for example, three articles in *Science, Technology & Human Values*: Martin (1993), Cozzens (1993), and Winner (1993), and also Carl Mitcham's review of *The Handbook of Science and Technology Studies* (1995) and Chubin (1992).

7. Contact information for the Loka Institute appears in note 2 above. The Council for Responsible Genetics' website is: <http://www.genewatch.org> and that of the Computer Professionals for Social Responsibility is: <http://www.cpsr.org>.

References

Chatfield, Rebecca, Henderson, Sarah Kuhn, and Muller, Michael, eds. 1998. *PDC 98: Proceedings of the Participatory Design Conference, Seattle, Washington, USA, 12–14 November 1998*. Palo Alto, CA: Computer Professionals for Social Responsibility.

Chubin, Daryl. 1992. "The Elusive Second 'S' in 'STS': Who's Zoomin' Who?" *Technoscience* (newsletter of the Society for the Social Studies of Science), vol. 5, no. 3 (fall), pp. 12–13.

Computers in Context: Joining Forces in Design. 1995. Third Decennial Conference Proceedings, Aarhus, Denmark, August 14–18, 1995. Aarhus: Dept. of Computer Science, Aarhus University.

Cozzens, Susan E. 1993. "Whose Movement? STS and Social Justice." *Science, Technology, & Human Values*, vol. 18, no. 3 (summer), pp. 275–77.

Fals-Borda, Orlanda, and Anisur Rahman, Muhammad. 1991. *Action and Knowledge: Breaking the Monopoly with Participatory Action-Research*. New York: Apex Press.

Fischer, Claude S. 1987. "Understanding Technology: An Agenda," Book review of *The Social Construction of Technological Systems*, eds. Wiebe Bijker et al. *Science*, vol. 238 (20 November), pp. 1152–53.

Guston, David. 1999. "Evaluating the First U.S. Consensus Conference: The Impact of the Citizens' Panel on Telecommunications and the Future of Democracy." *Science, Technology, & Human Values*, vol. 24, no. 4 (autumn), pp. 451–482.

Joss, Simon, and Durant, John, eds. 1995. *Public Participation in Science: The Role of Consensus Conferences in Europe*. London: Science Museum.

Leydesdorff, Loet, and Van den Besselaar, Peter. 1987. "What We Have Learned from the Amsterdam Science Shop." Pp. 135–60 in Stuart Blume, Joske Bunders, Loet Leydesdorff, and Richard Whitley, eds., *The Social Direction of the Public Sciences: Causes and Consequences of Co-operation Between Scientists and Non-Scientific Groups*. Dordrecht: D. Reidel.

Martin, Brian. 1993. "The Critique of Science Becomes Academic." *Science, Technology, & Human Values*, vol. 18, no. 2 (spring), pp. 247–59.

Meisner Rosen, Christine. 1989. Book review of *The City and Technology*, eds. Mark H. Rose and Joel A. Tarr. *Technology and Culture*, vol. 30, no. 4 (October), pp. 1070–72.

Mitcham, Carl. 1995. Book review of *Handbook of Science and Technology Studies*, eds. Jasanoff et al. *Science, Technology & Society* (the Lehigh University newsletter), no. 106 (winter), pp. 2–4.

Nyden, Philip, Figert, Anne, Shibley, Mark, and Barrows, Darryl, eds. 1997. *Building Community: Social Science in Action*. Thousand Oaks, CA: Pine Forge Press.

Park, Peter, Brydon-Miller, Mary, Hall, Budd, and Jackson, Ted, eds. 1993. *Voices of Change: Participatory Research in the United States and Canada*. Westport, CT: Bergin & Garvey.

Putnam, Robert D. 1996. "The Strange Disappearance of Civic America." *American Prospect*, no. 24 (winter), pp. 34–50.

Rogers, Everett M. 1983. *Diffusion of Innovations*, 3rd ed. New York: Free Press.

Sandberg, Åke, Broms, Gunnar, Grip, Arne, Sundström, Lars, Steen, Jesper, and Ullmark, Peter. 1992. *Technological Change and Co-Determination in Sweden*. Philadelphia: Temple University Press.

Schuler, Douglas, and Namioka, Aki, eds. 1993. *Participatory Design: Principles and Practices*. Hillsdale, NJ: Lawrence Erlbaum Associates.

Sclove, Richard E. 1995. *Democracy and Technology*. New York: Guilford Press (esp. chap. 11).

———. 1995. "Putting Science to Work in Communities." *Chronicle of Higher Education*, vol. 41, no. 29 (31 March), pp. B1–B3.

———. 1996. "Town Meetings on Technology." *Technology Review*, vol. 99, no. 5 (July), pp. 24–31.

Sclove, Richard E., Scammell, Madeleine L., and Holland, Breena. 1998. *Community-Based Research in the United States: An Introductory Reconnaissance, Including Twelve Organizational Case Studies and Comparison with the Dutch Science Shops and the Mainstream American Research System.* Amherst, MA: Loka Institute.

Winner, Langdon. 1993. "Upon Opening the Black Box and Finding It Empty: Social Constructivism and the Philosophy of Technology." *Science, Technology, & Human Values*, vol. 18, no. 3 (summer), pp. 362–78.

Zaal, Rolf, and Leydesdorff, Loet. 1987. "Amsterdam Science Shop and Its Influence on University Research: The Effects of Ten Years of Dealing with Non-Academic Questions." *Science and Public Policy*, vol. 14, no. 6 (December), pp. 310–16.

9

Gender:
The Missing Factor in STS

EULALIA PÉREZ SEDEÑO

Among the many societal factors inherent in the development of science and technology is the role of gender. However, scholars, especially feminist writers, have pointed to the ways that gender has often been peripheral to the social study of science and technology, even in some of the most constructivist-oriented work. It is Eulalia Pérez Sedeño's intent to rectify this omission, at least for this collection, first, by showing how women have largely been marginalized in the actual doing of science and technology; then, by calling for more attention to be paid to the contributions that women have, in fact, made to techno-science; and, finally, by arguing that for STS to be a viable educational and research program, it must routinely incorporate gender as an integral factor of analysis.

Pérez Sedeño, a professor of logic and philosophy of science at the Complutense University of Madrid, has published widely on the role of women in science, technology, and mathematics, especially with regard to matters of their education in Spain. She is currently editing a collection of essays on STS studies in the Hispanic-speaking world. Thus, she is particularly well-suited to raise questions regarding issues of gender in science and technology studies.

As we read this essay, it would be useful by way of comparison to consider how other authors in the collection have or have not incorporated gender in their visions of STS as a field of study. Is gender truly *the* missing factor, as Pérez Sedeño suggests?

There is absolutely no doubt that science and technology have substantially transformed, albeit often in erratic ways, the lives of human beings and the environment in which they live. In past centuries, such transformations were looked upon and lauded as evidence of Man's (*sic*) ability to dominate Nature, which, according to Francis Bacon's philosophy and the scientific revolution, was there to be subjugated and exploited by devices, artifacts, and inventions. Nature, for her part, responded by dispensing justice on the descriptive scientific theories and technology created for her, because she, and only she, is the judge who decides which theories and technology are acceptable.

During the last few decades, however, criticisms aired by some historians, sociologists, and philosophers of science, together with those of the ecological, feminist, and pacifist movements, have produced a type of reflection on science and technology that has put such a conceptualization on the rack. Although science and technology have different effects in different countries, which depends on differences of race, social class, and gender, we can safely assume that modern techno-science is largely a product of investigations performed by trained people who employ characteristic methods and techniques. Thus, techno-science is a body of knowledge and organized procedures, a means of solving problems. It is also a social and educational institution, one that requires material installations and a cultural resource, but, above all, it is a fundamental factor in human affairs.[1] That is to say, it is a system which has been formed by ideas, designers, artifacts, and end-users, among which there exists a synergistic relationship, one that exists in a specific socio-historical context.

Such a characterization has many advantages. Among others, it highlights the necessity of utilizing a multiplicity of disciplines and points of view, such as history, economics, psychology, sociology, so as to gain an accurate comprehension of techno-science. In addition, it presupposes a recognition that science and technology are not alien to human beings and that, as a consequence, they should be considered as having an interactive relationship with humankind. This means that there are implied ethical and political questions that cannot be put to one side when the time comes to determine whether techno-scientific practices are acceptable or not.

The above characterization had its origin in Science and Technology Studies (STS), but it is also a consequence of how STS has evolved. Nowadays, STS is a multidiscipline in which scholars examine science and technology issues and problems from different disciplinary perspectives that deal with cultural, political, social, and ethical questions about how science and technology should be developed and how they could be improved. When one refers to the various disciplines and areas that are covered by STS, one usually includes gender studies.[2] However, I believe that the inclusion of this perspective, in the majority of cases, is merely an attempt at being politically correct, and it is not a genuine recognition of what a gender approach to science and technology means or could mean. This can be seen if one examines the available STS literature in which there are relatively few studies that are at all sensitive to questions of gender. For example, in Bijker, Hughes, and Pinch (1987), one of the most important collections of essays on the social construction of technology, there is hardly any article that could be classified as dealing with that subject, and only one of the works was actually written by a woman. The same is also true of Bijker and Law (1992), an extension of themes developed in the volume mentioned earlier. Furthermore, when such studies do appear, as in the *Handbook of Science and Technology Studies* (Jasanoff et al., 1995), they appear to be peripheral asides that are not really integrated into STS.[3]

I wonder why such marginalization happens, given that one of the most important questions dealt with in STS is the interaction between the social and cultural factors in science and technology. One wonders even more when the function carried out by feminist thinking in a critical reflection on science and technology is of such importance nowadays.[4] It could seem impossible to ignore it, because it raises crucial questions for a comprehensive understanding of science and technology. As things stand presently, I believe that gender is a missing factor that has not sufficiently been taken into consideration by STS. The genuine inclusion of the gender approach is admittedly a challenge, for it would entail a further redefining of science and technology, and of what scientific and technological progress means. To do so, however, would also open the door to new issues, to new challenges, and to different solutions to the questions already posed, both in gender studies and STS.

Feminist Studies of Science and Technology

Feminist studies of science and of technology have not run chronologically in tandem, with a focus on technology having occurred only somewhat more recently. Nonetheless, the latter has followed the path of earlier feminist studies of science by posing similar questions. Even though feminism does not have a common singular stance with regard to techno-science, it does have a common base in maintaining that there is a gender bias in the majority of academic disciplines "that is expressed in particular claims and facilitated by disciplinary first principles" and that "women's experience is made invisible or distorted, as are gender relations" (Longino, 1999). As the majority of persons doing science and technology have been men, at least half of humanity has been ignored. That, in turn, has been due to, and reinforced by the fact that, traditionally, science and technology have considered themselves to be objective, neutral, and free of values. That is to say, they have assumed that "external" factors like gender have no place as a central element or concern in their makeup.

One of the general characteristics of feminism is the advancement of social and political issues that will lead to full equality for women. One result has been a focus on pedagogical issues and questions. Here a main objective was that more women should study science and technology, and get more involved in technoscientific activities. With this in mind, some feminist scholars conducted a study of how science and technology was taught at schools and higher educational institutions, and the contents of various curricula. The strategies used to encourage young girls and women to take up these subjects, and later to work in these areas, were very varied. Some focused on the content of the subjects, the selection of suitable texts, the inclusion of information not normally found in standard courses, and on the expectations that young girls and adolescents have from science, which normally condition their adult options. As well they focused on the conscious or unconscious consequences of expectations and attitudes toward female students of the teachers and professionals working in the fields of science and technology. Especially crucial is the need to provide female role models for young women who want to study, or dedicate their careers, to science.

To achieve this latter objective, several recent studies have plucked from oblivion the achievements of female scientists who had been inadvertently, or deliberately, banished from the traditional narratives of the history of science and technology either because of inherent biases, or narrow conceptualizations of the disciplines. There was, in fact, a rich array of women who had made significant contributions to the world of science and technology but who hitherto were not included in history books. The role of women in the origin and development of certain disciplines and allied subjects (such as botany, medicine, and programing) were examined, along with studies on the valuable contributions made by women in the development of techno-science (scientific and literary salons, scientific popularization, and the like). In the case of technology, this sort of recuperation was more difficult because of the systematic, legally permitted, nondisclosure of the names of women who had taken out patents in many countries. Another reason was the fact that standard works in the history of science and technology have largely passed the private sphere over—that is to say, the feminine one—in which the technologies used traditionally were, and continue to be, determined by a sexual division of labor. These studies have clearly shown that, throughout history, the number of women involved in scientific and technological disciplines is lower than that of males, but that the number is not as low as has been claimed. Nevertheless, their presence has been kept hidden away because of prejudice and outdated misconceptions about how the history of science and technology should be constituted.[5]

Studies of science and gender have also been concerned with identifying sexist and androcentric biases, which have shown up in particular theories and technoscientific practices throughout history. In particular, it has become obvious in those disciplines and practices directly related to human beings—the social sciences and biology—that such biases have occurred in all steps of scientific and technological practice, but especially in the selection and definition of problems, the planning of inquiry, and the collection and interpretation of data. Biology has been the subject of numerous analyses because of its pivotal role in the maintenance of the gender organization of society: starting with what biological beings *are*, through how they *should act* socially. In this regard, biological studies on the nature of human being have been shown to be flawed in experimental planning, in the gathering and interpretation of limited, if not contradictory, experimental data, and in unsustainable conclusions and fallacious arguments.[6]

Such criticisms readily show that older theories and practices were "bad science" in the traditional sense of the term, but it is also important to ask whether it is possible to detect that same bias in so-called good science. Analyses of the conceptual and metaphorical language used in science and technology have shown traces of sexist bias in many scientific and technological areas, which, not surprisingly, are often related to cultural ideals of masculinity, traditional rationalistic ideas, and alleged technoscientific objectivity. This also raised interesting questions regarding who learns what specific types of scientific knowledge, as well as questions regarding the neutrality and objectivity of technoscientific investigation.[7]

In the case of technology, these studies examined whether technology had, or had not, contributed to the emancipation of women. Many scholars believe Western technology has in the main been distinctly patriarchal and used to dominate women (Rowland, 1985; Corea, 1985; Corea et al., 1985; Mies, 1989; Grint and Gill, 1995). They claim that such technology, which is produced by men for use by women, can prove to be totally inappropriate for the needs of women, even to the point of being harmful as it incorporates masculine ideologies that dictate how women should live (Karpf, 1987, page 159).[8] At the same time, there are also techno-optimists. An example of this stance can be found in the work of Shulamith Firestone (1970). She saw contraceptive and reproductive technology as a mechanism of release for women from the tyranny of reproduction. Firestone believes the source of women's oppression is to be found in their biological being; thus, the possibility of putting a full-stop to biological motherhood, thanks to technology, was a milestone in the liberation of women.

The STS Perspective

With regard to such issues, an STS perspective incorporates a skeptical, albeit constructively critical, attitude, taking into account different points of view. For example, we can see that earlier and more recent sociocultural effects of reproductive technology are very different. Despite the fact that much reproductive technology is male designed, it is nonetheless the case that many women knowingly request technological treatment (contraceptive pills, epidural anaesthetic, amniocentesis, ultra-sound scans). Thus, at the same time that reproductive technology gives new possibilities to women, it also presents new challenges with regard to its appropriate adoption. STS can shed valuable light on this otherwise seemingly paradoxical situation.

Be it for good or evil, we all live in a world where technology is omnipresent. Technology is deeply integrated within all facets of our lives, public and private. The most sophisticated means of communication and transport, the chemical industries, genetically engineered food, domestic appliances, reproductive technology, all have had an effect on our society in different ways (polluting the environment, unequal development between the Third and Western worlds, incorporating women into the job market, or changing the very idea of family). Technology has also affected our private lives, for example, in the ways we communicate with others via the telephone and the internet. Technologies have also changed our resistance or immunity to certain illnesses and diseases by altering our metabolism, and they have established new forms of family relationships.

Unfortunately STS has primarily concentrated on the relationships of paid work and in the first stages of technological production, in effect sidelining fields such as reproduction, consumer needs, and nonremunerative productivity in the home. For example, STS has given relatively less consideration to many notable (including from an economic viewpoint) inventions for the private individual's daily use, such as disposable diapers and baby-feeding bottles. In contrast STS has more often concerned itself with "significant" high-profile artifacts and projects (cars, missiles, engines, and aircraft instruments), technologies that have been largely masculine orientated. STS also tends to concentrate on the public sphere, which, until recently, was frequently prohibited territory for women, while ignoring the private sphere almost completely. The latter is, of course, largely the feminine one, in which resides those technologies related to work that women carry out. This often neglected area continues to represent an analytical challenge that STS must take up.

This challenge presents STS with a fundamental problem of demarcation: what do we and do we not consider to be technoscientific. If we accept the traditional view of what constitutes techno-science, we would be largely excluding both those artifacts and practices invented by women, and those used primarily by them. In this traditional view, the ideological connection between masculinity and techno-science is also evident because masculinity is interpreted as synonymous with technical competence. And here competence means being able to use those technologies that men design and produce, most often those previously mentioned massive projects and industrial artifacts. In this way the association of masculinity and technology is constantly being strengthened, and it further underlines the fact that masculine power over technology is both a product and a reinforcement of his power in society.

It is also necessary to bear in mind the symbolic dimension of technology and how it impinges on our gender identity, which can be seen in many different spheres. For example, just recently in a press and television advertisement, there was a handsome father protectively holding a beautiful baby. It symbolized the security and the protection that the father was offering his family. As long as a man is the one portrayed as fixing a plug or the water tank under the watchful eye of his female partner, messages of inferiority and technical incompetence will be sent out to her. We transmit similar messages to our children when we give our sons presents of construction sets, machine kits, chemistry sets, or natural mineral collections, while we give our daughters Barbie Superstar dolls. The challenge is to break up this relationship between masculinity and technology in such a way that techno-science and technoscientific progress are redefined in ways that do not create dependence nor exclude anyone.[9]

Certainly one major concern within STS is to convey the idea that social and political decisions are inherent in the design and selection of any given technology, and thus that it is not merely a product of imperative technoscientific rationale. Because technology is the result of a series of specific decisions made by particular groups of people in specific places and times, with specific purposes in mind, its final form depends upon the distribution of power and the resources available inside each society.[10] Because each social group has its own particular interests and resources, the process of development often leads to conflicts arising from different opinions, for example, regarding what technical requisites should be established for a given artifact. We also should remember that the absence of influence will also shape science and technology, since a particular path will not be pursued. In a technological culture that is also a masculine culture, other less influential groups are frequently brushed aside. As a consequence, when analyzing the external factors that influence science and technology, one must, of necessity, include gender.

The Importance of Gender

The relationship between technology and gender is complex and depends on, among other things, how one views not only gender but also technology itself.

> Technology is more than a collection of physical objects or artifacts. It also fundamentally embodies a culture or set of social relations made up of certain sorts of knowledge, beliefs, desires and practices. Treating technology as a culture has enabled us to see the way in which technology is expressive of masculinity and how, in turn, men characteristically view themselves, in relation to these machines (Wajcman, 1991, p.149).

Thus, if we redefine technology in such a way that it includes the gender factor, the relationship between science, technology, and society takes on another form. If we analyze, in a critical way, certain "technological scenarios," we will come to the conclusion that technology is made up of an artifact/human-being interface which configures, accommodates, interferes with, and incorporates both human beings and their technologies into a network of social relationships, but without causally determining them. That is to say that technology is designed by human beings, men and women situated in specific economic, political, and historical circumstances, who, in part because they are of different sexes, have their own specific interests, and are in their own particular power situations.

It is also important to understand that while technologies are created with certain goals in mind, the end-users will often transform how they are used or use them to perform other tasks. It is not the case that a certain technology is created in the abstract and then put to another use (be it good or bad), for technology is always created by a "designer" who has a final aim in mind. End-users adopt technologies for specific purposes from the beginning, but they may build in improvements and extensions in such a way that the original is converted into a completely *different* technology and is thus unrecognizable from what was originally intended. This is what happened with the contraceptive pill, which started off by being a treatment to control the menstrual cycle of married women in order to help them to become pregnant. Thus, it developed as a family planning aid, but at the same time it became a means by which women could enjoy their sexuality without unwanted pregnancies, and an instrument which males could use to enjoy their own sexuality without unwelcome responsibilities. Or, take the case of the Internet, which was created by the military so that the "enemy" could not intercept or make use of classified information. However, it quickly developed into a participatory support technology in the hands of feminist groups, progressive political parties, NGO groups, and the like.

Science and technology are systems that contribute to shaping our lives, for they provide a framework in which we organize and carry out our actions. They can also frame our vision of social relationships and what it means to be a human being. Various branches of technology, especially those so-called technologies of gender and sex, can both modify existing hierarchical relations, and at the same time produce new ones according to gender (Haraway, 1991). Looked at from this perspective, gender is a technology too because it is "an organized system of management and control which produces and reproduces classifications and hierarchical distinctions between masculinity and femininity . . . ; it is a system of representations which assigns meaning and value to individuals in society, making them into either men or women" (Terry and Calvert, 1997, p. 6).

Contemporary society's tendency toward globalization not only includes the economy but also science and technology. The influences of this tendency are by no means uniform nor consistent for all countries, all social classes, or for all women. Thus, STS should not neglect the differential conditions of development nor the consequences that Western techno-science may bestow on marginalized groups, especially women and less developed countries.[11] In this regard, STS as an educational program should also deal with the minimum techno-scientific knowledge which everyone, irrespective of sex, class, or nationality, should have in order to make decisions. Above all, STS can, and should, play a fundamental role in making known what techno-scientific practices are indispensable for the better development of human beings, and they should point out policies that are best suited to that development.

For example, according to the United Nations' Report on Human Development, published in 1995,[12] most educational technologies never reach the majority of women. At least sixty million young girls all over the world have no access to primary education, while the number of young boys is this situation is forty million. More than two-thirds of the 960 million adult illiterates in the whole world are women. If we believe that education increases the capacity of a person to participate in society and improve their quality of life, including attaining better jobs and higher incomes, this illiteracy situation conditions the productive and reproductive lives of all those women, as do their scant incomes, not to mention their inability to participate in the decision-making process.

The case of medical technologies is also very revealing. According to the U.N., between 1970 and 1990, the life expectancy of women living in developing countries increased by nine years. However, an African woman lives approximately twenty years less than a woman in a developed country. In those less developed countries, a third of all women between the ages of fifteen and forty-four, suffer from illnesses related to pregnancy, giving birth, abortion, or problems with their reproductive organs. According to the World Health Organization, half a million women die every year due to complications experienced during pregnancy and in giving birth. Of these, 99 percent are from less developed countries. In these countries, the mortality rate for mothers is 420 deaths for every 100,000 children born alive, which is in contrast to thirty deaths for every 100,000 live births in developed countries.[13] According to the U.N., $140 million would need to be spent each year from now until 2005 to ensure adequate accessibility to medical services and family planning to redress this imbalance. Meanwhile, the developed Western world invests enormous amounts of money in assisted reproductive technologies, such as in vitro fertilization and intracytoplasmic sperm injection, even though this technology only benefits a select few. We cannot put aside these kind of issues, not if we want STS be an activist program that is politically and ethically honest and committed to education and research.

We all know that techno-science involves assumptions, acquisition of skills, norms of behavior, and compromises of values. That is to say, technology is not neutral, for while it responds to social necessities, it also creates them. It may solve problems, but it is also responsible for new ones. Increasingly, it gives rise to social, ethical, and political concerns out of which new technologies also emerge. One central feature in this value-laden nature of techno-science is the place of gender. Thus, gender studies needs to work closely with STS as they advance educational and political proposals that argue the necessity for a democratization of science and technology, one that includes full participation by *all* the world's citizens. Such democratization and participation cannot come about, however, if half the world's population remains excluded.[14]

Notes

1. Ziman (1984) offers a similar characterization for science.

2. Even though, as Keller 1995 suggests, we have no reason to identify gender with women—just as race is not to be identified as the black race—I will do so throughout this work, simply because gender studies has used this identification due to the fact that women have been historically and culturally tagged with gender.

3. Something similar occurs in journals. Even though the number of papers dealing with gender has increased, they are still small in overall number or are published as special issues, as was the case with "Gender Analysis and the History of Technology," in *Technology and Culture*, vol. 38 (January 1997), pp. 1–231, or else they are exiled to women studies' magazines.

4. Even though feminist thinking is by no means homogenous, I will adopt a very simple assumption—that is, one that starts from the fundamental idea that men and women have the same rights and capabilities. Although simplistic, this approach can lead far.

5. The bibliography regarding the role of women in science and technology is growing. With regard to the pedagogic aspect, see, for example, Alcalá (1998), Autumn (1990), and Kirkup and Smith Keller (1992). For comments on institutional barriers and discriminatory regulations, see Fox (1995), Kirkup and Smith Keller (1992), Pérez Sedeño (1995a), (1995b), (1996), (1997) and (1998), Sonnert and Holton (1995), and Rossiter (1982) and (1995). One of the latest overviews with an extensive bibliography, can be found in Kohlstedt and Longino (1997).

6. Works on this subject are also very numerous, for example, Bleier (1979 and 1984), Fausto-Sterling (1992), especially chapter 8, Hubbard (1992), Longino (1990, chapters 6 and 7, and 1995), Longino and Doell (1983), and Sayers 1982.

7. In the last ten years there has been a plethora of epistemological studies of this type, but they have fundamentally dealt with science. For a panoramic overview, see, for example, Harding (1986), Longino (1999), Nelson and Nelson (1996) and González & Pérez Sedeño (1998).

8. This position certainly seems to advocate for an essentialism which, in my opinion, is not very well founded at all: anthropological studies of various cultures show that there are no behaviors or meanings universally associated with women (nor with men), but they are constructed socially and historically. On the other hand, I find that some stances, such as ecofeminism, sound out of tune when they regard the essence of femininity as being as close as possible to nature and especially, with regard to biology, and reproductive capacity—power, they say—which is exactly what has been used to keep women subordinate throughout history. However, there are numerous authors who defend this thesis. See, for example, Merchant (1980).

9. I realize that this ideological supposition is strong and deeply rooted in our culture and produces certain effects; there is also a similar ideological bond between race and techno-science, as well as between class and techno-science.

10. Although, as is pointed out by Cowan (1997), who and how many people go to make up those groups are different in science and in technology, and they will also vary according to the epoch.

11. Women constitute a nonquantitative social minority in the same way as do other social minorities categorized by race, sexual orientation, or marginalized urban status. The same can be said of less developed countries.

12. Data are taken from "Women of the Third World," published by NGO *Medicus Mundi*, in March 1996. I thank Eduardo de Bustos Pérez for providing me access to this information.

13. According to the Spanish National Statistics Institute, 2.9 mothers die for every 100,000 live births in Spain.

14. The present work was made possible, in part, thanks to financial help given by the CICYT of the Spanish government, Research Project, PB95-0125-C06-03. I would like to thank Stephen Cutcliffe, Marta I. González, and Paloma Alcalá very much for their useful and penetrating comments. Stephen Cutcliffe, especially, has proved indispensable for an understanding of this paper.

References

Alcalá, Cortijo, and Soledad, Paloma. 1998. "Sobre los ingenios femeninos." In Alcalá, Soledad, and Pérez Sedeño, eds.

Alcalá, Cortijo, Soledad, Paloma, and Pérez Sedeño, Eulalia, eds. 1998. *Actas del I Congreso Multidisciplinar 'Ciencia y género.'* Madrid: Universidad Complutense.

Bijker, Wiebe, and Law, John. 1992. *Shaping Technology / Building Society: Studies in Sociotechnological Change*. Cambridge, MA: MIT Press.

Bijker, Wiebe E., Hughes, Thomas P., and Pinch, Trevor, eds. 1987. *The Social Construction of Technological Systems: New Directions in the Sociology and History of Technology*. Cambridge, MA: MIT Press.

Bleier, Ruth. 1979. "Social and Political Bias in Science: An Examination of Animal Studies and Their Generalization to Human Behaviors and Evolution." Pp. 49–69 in Ruth Hubbard and Marian Lowe, eds., *Genes and Gender II*. New York: Gordian Press.

———. 1984. *Science and Gender: A Critique of Biology and Its Theories on Women*. New York: Pergamon Press.

Corea, Gena. 1985. *The Mother Machine: Reproductive Technologies from Artificial Insemination to Artificial Wombs*. New York: Harper and Row.

Corea, Gena, Klein, Renate Duelli, Hanmer, Jalna, Holmes, Helen B., Hoskins, Betty, Kishwar, Madhu, Raymond, Janice, Rowland, Robyn, and Steinbacher, Roberta, eds. 1985. *Man-Made Woman: How New Reproductive Technologies Affect Women*. Bloomington, IN: Indiana University Press.

Cowan, Ruth Schwartz. 1997. "Domestic Technologies: Cinderella and the Engineers." *Women's Studies International Forum*, vol. 20, no. 3, pp. 361–71.

Fausto-Sterling, Anne. 1992. *Myths of Gender*. New York: Basic Books.

Firestone, Shulamith. 1970. *The Dialectic of Sex*. New York: William Morrow.

Fox, Mary Frank. 1995. "Women and Scientific Careers." Pp. 205–28 in Jasanoff et al., eds., 1995.

González García, Isabel, Marta, and Pérez Sedeño, Eulalia. 1998. "Ciencia, tecnología y género." *Revista Iberoamericana de Educación*.

Grint, Keith, and Gill, Rosalind, eds. 1995. *The Gender-Technology Relation: Contemporary Theory and Research*. London: Taylor and Francis.

Haraway, Donna. 1991. *Simians, Cyborgs, Women*. London: Routledge.

Harding, Sandra. 1986. *The Science Question in Feminism*. Ithaca, NY: Cornell University Press.

Hubbard, Ruth. 1992. *The Politics of Women's Biology*. New Brunswick, NJ: Rutgers University Press.

Jasanoff, Sheila, Markle, Gerald E., Petersen, James C., and Pinch, Trevor, eds. 1995. *Handbook of Science and Technology Studies*. Thousand Oaks, CA: Sage.

Karpf, A. 1987. "Recent Feminist Approaches to Women and Technology." In McNeil, ed., 1987.

Keller, Evelyn Fox. 1995. "The Origin, History, and Politics of the Subject Called 'Gender and Science': A First Person Account." Pp. 80–94 in Jasanoff et al., eds., 1995.

Kirkup, Gill, and Smith Keller, Laurie, eds. 1992. *Inventing Women: Science, Technology, and Gender*. Cambridge: Polity.

Kohlstedt, Sally Gregory, and Longino, Helen, eds. 1997. *Women, Gender, and Science. New Directions. Osiris*, vol. 12.

Longino, Helen. 1990. *Science as Social Knowledge*. Princeton, NJ: Princeton University Press.

———. 1995. "Knowledge, Bodies, and Values: Reproductive Technologies and Their Scientific Context." Pp. 195–210 in Andrew Feenberg and Alistair Hannay, eds., *Technology and the Politics of Knowledge*. Bloomington, IN: Indiana University Press.

———. 1999. "Feminist Epistemology." In John Greco and Ernest Sosa, eds., *Blackwell Guide to Epistemology*. Malden, MA: Blackwell.

Longino, Helen, and Doell, Ruth. 1983. "Body, Bias and Behaviour." *Signs*, vol. 9, no. 2 (winter), pp. 206–27.

Merchant, Carolyn. 1980. *The Death of Nature: Women, Ecology and the Scientific Revolution*. New York: Harper and Row.

McNeil, Maureen, ed. 1987. *Gender and Expertise*. London: Free Association Books.

Mies, Maria. 1987. "Why Do We Need All This? A Call Against Genetic Engineering and Reproductive Technology." Pp. 34–47 in Patricia Spallone and Deborah Lynn Steinberg, eds., *Made to Order: The Myth of Reproductive and Genetic Progress*. New York: Pergamon Press.

Nelson, Lynn Hankinson, and Nelson, Jack, eds. 1996. *Feminism, Science and Philosophy of Science*. Boston: Kluwer.

Pérez Sedeño, Eulalia. 1995a. "Scientific Academic Careers of Women in Spain: History and Facts," paper presented in the *VIth ILS Conference*, Frankfort, KY.

———. 1995b. "La sindrome de l'*Snark* i altres històries." *Quaderns del Observatori de la comunicació científica*, no. 1, pp. 58–70.

———. 1996. "Family *versus* Career in Women Mathematicians." Pp. 211–19 in *Proceedings of the Seventh European Women in Mathematics (EWM)*, Copenhague/Madrid.

———. 1997. "Decisiones injustas, decisiones innecesarias." Pp. 21–37 in *Actas del II Congreso de Coeducación en matemáticas*, Madrid, Sociedad "Ada Lovelace" para la Coeducación en Matemáticas.

———. 1998. "Las amistades peligrosas." Pp. 27–56 in Amparo Gómez et al., eds., *La construcción social de lo femenino*, La Laguna, Edic. de la Universidad de La Laguna.

Rossiter, Margaret. 1982. *Women Scientists in America: Struggles and Strategies to 1940*. Baltimore: Johns Hopkins University Press.

———. 1995. *Women Scientists in America: Before Affirmative Action*. Baltimore: Johns Hopkins University Press.

Rowland, Robin. 1985. "Motherhood, Patriarchal Power, Alienation and the Issue of 'Choice' in Sex Preselection." Pp. 74–87 in Gena Corea et al., eds., 1985.

Sayers, Janet. 1982. *Biological Politics: Feminist and Anti-Feminist Perspectives*. New York: Tavistock.

Sonnert, Gerhard, and Holton, Gerald. 1995. *Who Succeeds in Science?* New Brunswick, NJ: Rutgers University Press.

Stanley, Autumn. 1990. "The Patent Office Clerk as Conjurer: The Vanishing Lady Trick in a XIXth Century Historical Source." Pp. 118–36 in B. Dry Gulsky, ed., *Women, Work, and Technology*. Ann Arbor: University of Michigan Press.

Terry, Jennifer, and Calvert, Melodie, eds. 1997. *Processed Lives. Gender and Technology in Everyday Life*. New York: Routledge.

Wajcman, Judy. 1991. *Feminism Confronts Technology*. University Park, PA: Pennsylvania State University Press.

———. 1995. "Feminist Theories of Technologies." Pp. 189–204 in Jasanoff et al., eds., 1995.

Ziman, John. 1984. *An Introduction to Science Studies. The Philosophical and Social Aspects of Science and Technology*. Cambridge: Cambridge University Press.

10

Postmodern Production and STS Studies: A Revolution Ignored

WILHELM E. FUDPUCKER, S.J.

As an external observer and sometime critic of the STS field, Wilhelm Fudpucker's contribution to the current collection provides a useful counterpoint to the nine more "insider" essays. Like Volti, Fudpucker focuses on a particular issue closely related to work; but like Pérez Sedeño, he sees general implications in the failure of STS to address the issue he raises.

Fudpucker is a physicist, philosopher, and sometime epic poet, and the recently retired director of the Massenheim Institute, a small think tank in Germany. He was at one time a student of Pierre Teilhard de Chardin and a follower of Emile du Bois-Reymond. His *Shepherd of Becoming* (1978) was an important early work relating technology to religion, as was his essay "Through Christian Technology to Technological Christianity," in Mitcham and Grote (1984). Since then he has focused on more secular topics, including science-technology-society relations. Indeed, the present essay, which was originally presented as a talk to a small group of international STS students from across Europe visiting the Massenheim Institute on the occasion of his 75th birthday, calls for the recognition of important changes that STS scholars no longer seem to have the imaginative interest to appreciate. (The lecture form of address has been retained here, which also reflects the challenging character of Fudpucker's argumentative style. We thank Professor Steven Goldman for calling our attention to this talk, and for providing the translation from its transcription.)

Key to the argument for an STS studies that takes account of postmodern forms of production is, perhaps somewhat ironically, a restatement of the idea of technological determinism. As Fudpucker says, "Technological innovation remains a principal driver of the shift from mass to postmodern production." At the same time, Fudpucker makes favorable mention of Wiebe Bijker's social constructivism. To what extent does he thus bridge the gap between the determinists and the constructivists? In reading this argument we also need to ask ourselves to what extent the driver is technology and to what extent it may be economic enterprise. Finally, it would useful to consider in what ways the new forms of production—rapid new product introduction, production to order, individualization at little or no extra cost—may be affecting educational as well as industrial institutions.

The Challenge to STS

On this—for me—auspicious occasion, I want to challenge you, precisely *you* young scholars, to overthrow the STS studies paradigm that your teachers boldly created thirty-five years ago, but which you know in your own lives no longer to be relevant. What your teachers judge to be superficial and regrettable features of contemporary popular culture, you know to be deep structural changes in the ways that wealth and power are being created and distributed worldwide. Your teachers continue to interpret relationships among ideas, machines, and values as the evolving consequences of three developments: the steam power-based industrial revolution, the invention of the modern industrial corporation, and the exploitation of science and technology by national governments in pursuit of their economic, political, and military agendas (and, as we know all too well here, political and military adventures).

The truth that your own experience reveals, however, whether you are consciously aware of it or not, is that each of these developments has over the past decade been superseded in at least three ways.

First, the enterprise-based, manufacturing-centered mass production technologies of the industrial revolution are no longer at the center of marketplace value. They have been displaced by new technologies that have enabled the emergence of distributed global "interprises"—networks of interconnected enterprises—whose goal is the collaborative production, not of products or even of familiar services, but of "solutions": customer-specific integrations of goods, information, and new kinds of services.

Second, and concurrently, the modern industrial corporation—vertically integrated, hierarchically organized, centrally administered—has proven to be increasingly dysfunctional. The most valuable markets today are rapidly and unpredictably changing niche, as opposed to mass, markets for individualized solutions. The command-and-control management philosophy and centralized authority structure of the modern corporation is as incompatible with competing effectively in these markets as large-scale, specialized production facilities designed to benefit from economies of scale.

The value of being able to respond quickly to constantly changing customer opportunities is forcing companies to manage their operations from the outside (marketplace) in, rather than from the inside out. But managing from the outside in requires loosening control, decentralizing decision-making, and distributing authority to a workforce that is knowledgeable, motivated to develop new solutions, and empowered to change the status quo. Managing from the inside out, on the other hand—that is, optimizing internal procedures—is precisely the objective of all inherited management science "wisdom." It is the long shadow cast by Frederick Winslow Taylor, Henry Ford, and Alfred Sloan.[1] But even the shadows of giants have limits!

Third, and finally, national governments have clearly lost the ability to determine by their policies the economic security of their citizenry. Increasingly, local prosperity, in developed no less than in undeveloped nations, is in the hands of globally distributed, multi-enterprise production networks able to exploit new communication- and information-processing technologies to link facilities and people opportunistically into functional business operations regardless of their physical location. In the process, national boundaries and policies are not so much evaded as ignored.

How has the STS studies community responded to these changes? With indifference, as if what were happening under their ordinarily very astute noses was not worthy of a scholarly response—precisely the attitude, back in the 1960s, of older historians, philosophers, and sociologists to the then-revolutionary challenge to expose the role of technical knowledge and things in human affairs. In short, STS studies have become narcissistic. STS is self-conscious about the academic integrity of its scholarship and looks for approval to an audience of academic STS scholars. No one is looking outside the received STS framework—looking, for example, at business news—for signs of developments that would require a transformation of that framework. There are, nevertheless, abundant signs in the world of commerce that the time has come for a new STS to be born. These new STS studies will reflect the forms in which new forces of production now rising to dominance are being institutionalized.

Reinventing STS Studies

To discover these new forms, however, requires turning away from STS studies themselves—away from the STS literature, away from familiar STS resources—and rediscovering the world in which wealth and power are created and exploited, as the STS pioneers had to do forty years ago. The long-standing Western academic bias against commerce and its institutions seems, ironically and paradoxically, to have infected even STS, which was born by taking things and their production seriously! That bias must again be overcome by you students, or STS will become increasingly impotent to illuminate the ways in which science, technology, and society act on one another as we enter the twenty-first century.

The changes in commerce to which I have been referring in fact have profound implications for STS studies. Necessarily, the study of science-technology-society interrelationships has reflected, until now, the continuing dominance of an industrial revolution-based system for the mass production of standardized goods and services. In the course of the last decade, however, a new "logic" for the integration of people, organizational design, and technology into a system for producing new kinds of goods and services and in new ways has risen to dominance. This post-mass production system was first given a distinctive name in a 1991 report on the future of U.S. industry funded by the Department of Defense and published by the Iacocca Institute at Lehigh University (Goldman and Preiss, 1991).[2] The report, *21st Century Manufacturing Enterprise Strategy: An Industry-Led View*, offered a detailed vision of what it called "agile" manufacturing and of the revolutionary organizational changes that would be required to create effective "agile" enterprises:

- Targeting high-value, rapidly changing *niche* markets;
- Treating manufacturing as a service supporting the sale of information- and services-intensive "solutions";

- Forming mutually beneficial, opportunity-driven, interenterprise relationships linking producers, suppliers,
  customers, and partners (even competitors!) into "virtual
  organizations."

Here too in Germany there was a recognition that new technologies enabled fundamental change in the structure of industrial
organizations. Some of what is called in America, and today worldwide, the "agile enterprise," was described by Hans-Jürgen Warnecke, director of the national network of Fraunhofer Institutes for
applied research and development, in his book *The Fractal Corporation* (1993): replacing rigid, centralized managerial control with
many autonomous yet coordinated work units. Similarly, agility
echoes features of the "lean" manufacturing model popularized in
the very influential *The Machine That Changed the World* (Womack, Jones, and Roos, 1991), especially the notions of efficiency and
a system-level treatment of the production of goods and services.[3]

For me—and, I hope, for you, too—calling this new system of
production "postmodern" is more revealing than "agile," "fractal," or
"lean." These names are abstract, technical, and objective-sounding,
while "postmodern" directly links production to a cultural context
with which it seemed to me to have strong affinities. I will return to
this theme before concluding this chapter. Whatever it is called, however, the new system of production unquestionably is displacing—better, is *marginalizing* in terms of market value—the mass production-
based system that dominated twentieth-century economic, social, and
political life. In the process, the form of the relationships among science, technology, and social institutions and values is changing, too.
That is the revolution that strikes at the heart of STS studies.

The challenge of the new postmodern production revolution to
STS, then, is to apply the intellectual orientation of interdisciplinary STS scholarship to exposing the ways in which emerging new
commercial realities are being institutionalized. How are science
and technology being harnessed to the creation of wealth and power
in this new production system, and how are research and innovation being transformed by new ways of pursuing wealth and power?
The goal should be to identify new drivers of change, and new forms
of interaction among scientific knowledge, technological innovation,
and (corporate, governmental, and social) institutional agendas,
and to provide some understanding of how these are already affecting society and will for the foreseeable future continue to do so.

The challenge is not for STS scholars to predict the social consequences of new technologies—for example, biotechnologies, nanotechnologies, quantum computing, or Internet-based commerce. That territory is well staked out by futurologists and certain business and science fiction writers, and is not always distinguishable from scholarship![4] What should be a matter of the deepest concern is that STS scholars—devoted to understanding the mutual relationships among science, technology, and society—should address the significance of commercial events around them that are causing fundamental changes in those relationships. One cause for concern in failing to do so is the implied assumption that STS relationships will always be determined by the evolving consequences of the industrial revolution, the rise of the modern corporation, and familiar forms of government involvement with science, technology, and industry. Such an assumption denies the manifestly historical, and therefore temporally and culturally delimited, character of "modernist" science-technology-society relationships.

Marketplace Change

What is of central importance for STS studies is not the technologies on which a production system depends, but Jacques Ellul's provocative notion of *la technique*, the coordinated organizational, social, and cultural values that alone make possible and sustain the exploitation of technologies as a means for creating and selectively distributing wealth, and for acquiring and exercising power (Ellul, 1954/1964).[5] Indeed, one of the most surprising revelations for many corporate executives who have been transforming their companies into postmodern enterprises is the recognition that how resources are organized is vastly more important than the resources themselves; that in the real world of industrial competition, technique trumps technology! The same human and physical resources, organized in different ways, will be capable not merely of different levels of productivity, but of creating vastly different kinds of value.[6] This fact, now widely acknowledged as such, was largely obscured by the complacency with which it was assumed that the inherited corporate organizational structure and interenterprise relationships were givens, not open to fundamental change.

Two marketplace developments undermined the profitability of the mass production system and the viability of the modern corporation: (1) the unprecedented rapid pace and unpredictability of marketplace change; and (2) the shift in the center of marketplace value from standardized products and services sold in single event transactions to individualized solutions sold as one event in a continuing relationship.

Every industry has its own rate at which its products and services are introduced, modified, and then superseded by newer products and services. Since the early 1980s the "clockspeed" of every industry has increased dramatically and continues to accelerate.[7] Automobile models used to be replaced with truly new models every five to seven years. Today, it is a significant competitive asset to be able to bring out new models in less than three years. In consumer electronics, computer hardware and software, photographic equipment, and cellular telephones, as in agricultural equipment, computer-controlled machine tools, and even financial services, the ability to introduce new products and new services quickly and often has become a decisive competitive weapon (Davis and Meyer, 1998; Stalk and Hout, 1990).

The technologies that enabled companies to offer a constant flow of new models and new products[8] are those associated with computer-controlled machine tools. Together with increasingly sophisticated data and information processing capabilities extended to sales, marketing, suppliers, and inventory management, many companies discovered that they could now produce goods and services to order instead of following often inaccurate forecasts. By now, everyone has heard many times over how Dell Computers has sustained its astonishing rate of growth and profitability by building millions of computers a year, each to individual customer order (Dell and Fredman, 1999).

But the Dell story is just one instance of thousands in which former mass-production-to-forecast systems have given way to high volume production-to-order systems. Furthermore, the same information technologies that allow companies to offer these individualized products and services come at no greater cost to the producer than standardized products and services. Each Dell PC is configured with the hardware and preloaded software of each customer's choice, and this is true for Gateway, IBM, and now Compaq as well.

Solutions, Not Products

The combination of these three capabilities—rapid new product introduction, production to order, individualization at little or no extra cost—together with the expansion of mass production manufacturing in economically undeveloped ("poor") countries with labor costs a small fraction of those in developed ("rich") countries, sounded the death knell for the continued profitability of mass production of goods and services in "rich" countries. The question that remained open was how these new production capabilities should be used to create profitable companies in the "rich" countries.

The "correct" strategy for responding to east Asian competition was to couple the three production system capabilities just described to a second marketplace development: the emergence of solutions as the center of marketplace value. In place of standard, general purpose goods and services, solutions were special purpose goods and services that created value for buyers.[9] For industrial customers (and most manufacturing companies sell their output to other companies, not to consumers), solutions are valued because they help create value (often subjective!) for their customers.

In place of general purpose sneakers, for example, special purpose sport shoes are tailored to the characteristics of a specific sport and even customized to the individual wearer. In place of a general purpose bicycle, vendors offer dirt bikes (with or without front or rear suspension), road bikes, racing bikes, touring bikes, hybrid "city" bikes, and "retro" cruisers: nostalgia is a very high-value niche market. Buyers can choose a frame made out of various steel alloys, aluminum, titanium, thermoplastics, or carbon fiber composites. They can select among variations on the traditional diamond "men's" frame, cantilever frames, upright posture, racing posture, or a recumbent (long or short wheelbase, above- or below-the-seat steering). They can have wire wheels, spoke wheels, or disc wheels, brake pads, or disc brakes. . . . The list of options is very long, and getting longer; the items continue to change very rapidly. In effect, every bicycle producer today is a systems integrator for niche markets, some of which are high-volume and some low, but the important fact for business is that the value of the global bicycle market is vastly higher than it was in the 1960s and 1970s. These are typical, not exceptional, examples. More and more manufacturing companies announce in their advertising that they are not manufacturing companies anymore, but solutions companies. Xerox, for one, does not sell copiers, though it does manufacture them. It sells distinct document processing solutions.

Wants Versus Needs

On the consumer side, a fundamental reality that obviates classical industrial production system economics is that the economies of "rich" countries are driven by what people want, not what they need. There is, in classical terms, no rational relationship between cost and price—no "fair" price derivable from the cost of materials, labor, and production capital—for what people want. There is no classically rational price for information and services, either. At first, companies charge a high price for technology novelties: ultralight cellular telephones, for example, or compact DirectTV satellite dishes. Very quickly, however, the physical product becomes a commodity that is almost given away to sell service contracts. All the continuing profits are in the services, not in selling the hardware.

That is why it was a mistake in the late 1980s for Western industry to apply the new flexible, computer-based production technologies they were installing to compete with the Japanese for manufacturing excellence. The value of manufacturing as a fraction of the sales price of solutions is small and getting smaller (which goes a long way toward explaining the continued doldrums in which the Japanese economy is adrift: Japanese companies have not yet figured out how to sell solutions).

The correct strategy for postmodern production remains moving from selling what you produce to selling what customers want. In the most profitable markets, what customers want is some combination of physical product, information, and services tailored to their evolving individual desires or requirements. Customers want solutions, and more and more companies are offering them. Notice, however, that for a company to be able to sell solutions, it must have capabilities that are fundamentally different from the capabilities that were characteristic of mass-production-era companies.

Cooperate to Compete

A postmodern producer must be able to create a stream of constantly changing models and products that it can integrate into individualized information- and services-rich solutions, and deliver them to order with much shorter lead times than ever before. To meet these objectives, companies across all industries are being forced to alter their fundamental structures and modes of operation, and these changes in organization and mode of operation are now beginning to affect health care, education, and even government agencies. Suddenly, cooperative relationships, within a company and between companies, are the key to competitive success (Goldman, Preiss, and Nagel, 1996).

Internally, and against the grain of traditionally adversarial labor-management relationships, the theme of cooperation manifests itself in three ways:

- Replacing the command-and-control philosophy of management characteristic of the industrial corporation as originally conceived, with a motivation-and-support philosophy—that is, the task of management is to motivate the entire workforce to achieve the success of the company by working, as the Americans love to say, "smarter, faster, cheaper, better"; and to support the workforce as its members display initiative and creativity by providing the necessary resources to bring proposed new solutions into reality.

- Organizing the company around workforce teams that break traditional functional authority boundaries—thus, crossfunctional solution design teams would include design department personnel, manufacturing personnel, sales, service, marketing, inventory management, purchasing—and, ideally, are self-managed and empowered to make relevant decisions at the operational level.

- Creating a truly open information and communication environment, opening lateral as well as vertical channels of communication and providing the workforce with traditionally proprietary information about the company's finances, the market value of operations, about customers and suppliers, about what is being done, how it is being done, and who is doing it.

Externally, cooperation manifests itself in newly interactive, and explicitly mutually beneficial, relationships with customers, suppliers, and partnering companies, including forming opportunistic partnerships with direct competitors. Today, the need to create new solutions continually and to bring them to market very quickly demands dynamic, collaborative relationships with suppliers, customers, and partners, not static sequential relationships. Companies find that it is in their own best interest to help their suppliers improve their products and processes in order to reduce supplier lead time and cost of production—perhaps to improve quality as well. In the process, the supplier benefits and so does their customer. The reverse is true as well, of course—that is, suppliers benefiting by teaching their customers how to do things differently in order for the supplier to provide more value to them. Chrysler Corporation—now DaimlerChrysler, of course—has been reaping more than a billion dollars in savings a year from a program in which it shares fifty-fifty with suppliers any cost reductions realized by Chrysler as a result of supplier initiatives.

In postmodern production one must think of numerous, constantly changing, mutually value-adding value circles. In this context, the notion of electronically linking physical and human resources that happen to be distributed among a group of companies—some traditionally thought of as suppliers, some as partners or competitors, others as customers—into a single virtual company in order to develop, produce, and market a proposed new solution makes a great deal of sense. To create solutions, companies must develop close, so-called proactive relations with customers, in order to understand what it is their customers will value. It is too late to wait until customers tell their suppliers what they want; the supplier has to anticipate the customer's evolving requirements, what the customer will recognize as valuable when they have it handed to them. Remember, no one asked for music compact discs, automated teller machines, or self-service gasoline stations!

Finally, creating solutions requires a workforce that is informed enough to recognize new customer opportunities, and both knowledgeable enough and motivated enough to respond to these opportunities innovatively, by creating new solutions that may require linking their company's resources in other companies. Postmodern production thus puts a premium on a knowledgeable and entrepreneurial workforce even as the workplace becomes increasingly unstable. Companies are responding to this challenge by installing complex knowledge management systems and by changing employee recruiting criteria and education and training methodologies. But the real test will be the ability to create company cultures in which employees with fragile job security will be motivated to "think like an owner" and work harder and smarter for the greater good of the company first.

STS in the Emerging Postmodern Industrial Era

It should be obvious that any institutionalized system for producing goods and services is directly linked to the ways in which wealth and power are created and distributed in a society. Since the 1960s, STS studies have helped us to understand some of the ways in which social, political, and economic institutions in the modern West have been linked to science and technology. They have done so by showing that, for the last hundred years especially, these linkages have involved reciprocal relationships among industrial corporations, the military, and economic agendas of national governments, and a broad range of increasingly sophisticated science-fed technologies of mass production.[10]

Today, a decisive shift can be discerned in the terms of the STS equation. Driven by the rapidity of marketplace change and the demand for a continuing stream of innovative solutions, the industrial corporation is undergoing deep structural change, though still, of course, in pursuit of maximum profit. Internally, the pressing need to decentralize decision-making processes and create an open information and knowledge environment entails a new management philosophy and new management-workforce relationships. Externally, the traditional well-defined enterprise with clear boundaries is giving way to what some call *interprises*, companies with fuzzy boundaries that have integrated their business, information, and production systems with those of suppliers, partners, and customers. It is almost commonplace today to find personnel from these classes of companies working intermingled in their various facilities. As companies organize to integrate business processes, often on a global basis, they create opportunistic virtual enterprises that follow market opportunities.

The ability of national governments to impose self-serving economic agendas on companies operating within their borders has been severely eroded by the emergence of global production networks among groups of collaborating companies. These networks are capable of producing goods and services for any local market by integrating the most efficient required production resources wherever they are located, anywhere in the world. Can even such a virtual (economic) entity as the European Union and its euro virtual currency provide a controlled business environment in the face of such a capability?

Technological innovation remains a principal driver of the shift from mass to postmodern production by driving rapid marketplace change and making possible cost-effective mass customization and the creation of apparently endless streams of solutions. In the mass production-dominated industrial era, benefiting from innovation was very capital-intensive. Companies had to build large-scale dedicated production facilities to begin generating profits to invest in growth. In the postmodern industrial era, solutions aimed at niche markets allow small and medium-scale venture capital firms to begin the profit-generation process.

Earlier, I said that there were important affinities between cultural postmodernism and the "agile" production systems now rising to dominance. I say this because the rationality of the familiar industrial corporation was recognizably the same as the rationality associated with modern science; hence the discipline of management science, which presumed that there was a logic to business decision-making. Business practices could be analyzed, modeled, and optimized as quasi-closed systems, just the way scientists treated domains of nature. Today, business is suddenly open-ended, unpredictable, constantly adapting to changing customer values driven by subjective judgments of worth. To the extent that business is logical today, it is the logic of the fashion and entertainment industries, putting a high value on constant novelty, independent of existential need or functionality. Control is an illusion; self-organization of personnel is the ideal. Neural net analogies are taken up as business models—specifying inputs and desired outputs, with managers allowing the workforce to decide how to achieve them—along with models from chaos theory and even quantum mechanics! Boundaries are illusory, too, as companies move from acting as Machiavellian city-states to nodes in constantly shifting networks of mutually beneficial relationships, and as businesses become engaged with the professional and personal lifestyles of commercial customers and consumers. Perhaps totalitarianism has shifted from the political to the commercial sphere?

In any event, the themes of social construction of knowledge and of technology—as expressed by Wiebe Bijker (1995), for example—have a new resonance with the world of commerce. In a post-mass production postmodern industrial corporation (but definitely not postcapitalist!) era, what is the "logic" by which new forms of commerce are being constructed today, integrating knowledge, technology, and social institutions into new means for creating and distributing wealth?

These are exciting times for reinventing STS studies, for *you* young people to reinvent STS studies. As the first postmodernists from birth, seize the initiative! Do not wait for your teachers—expert as they are in the entrenched paradigm—to open *your* eyes to the new ways that personal and social values are influencing and being influenced by scientific research and technological innovation. Read current STS studies for insight, but then use your own postmodernist sensibilities to identify the features of the world as you experience it that illuminate emerging configurations among ideas, machines, and values. The old paradigm helped us to see relationships among science, technology, and society that had been ignored or invisible before. The new paradigm is waiting to be born; the job of delivering it is yours!

Notes

1. Kanigel (1997) is an excellent biography. Halberstam (1986) exposes the management philosophy of the U.S. auto industry through the 1970s, contrasting it with the then-rising Nissan and Toyota companies. The classic study of Sloan-ism remains Peter Drucker's 1946 study of General Motors (1946).

2. The principal investigators of the DoD-funded project summarized in the report were Roger N. Nagel and Rick Dove. See also Goldman, Preiss, and Nagel (1995) and (1996).

3. See also the more recent volume, Womack and Jones (1996), and for a critique of the human impact of lean, see Harrison (1994), and Derber (1998).

4. For example, see the website of the World Futures Society, www.wfs.org. Examples of prophetic technology-creating-business books are: Gilder (1989); Drexler (1992); and Moravec (1999). On the science fiction side: Brunner (1975); Gibson (1984) and (1987); and Stephenson (1992) and (1996) are proving remarkably prescient.

5. I note that the Germans have never seen fit to translate this book, perhaps in the mistaken belief that the work of Helmut Schelsky is an adequate substitution.

6. A pioneering and still valuable study of the relationship between organizational structure and individual performance is Jacques (1991).

7. I first heard this "clockspeed" metaphor used generically in a presentation by Professor Charles Fine of MIT in 1996. It is commonplace today.

8. For example, in one thirteen-month period in the mid-1980s, Honda brought out eighty-five new models of motorcycles in a successful defense of its status as the world's leading motorcycle manufacturer against a bid by Yamaha to move up from number two.

9. See Goldman, Preiss, and Nagel (1995) for a full discussion of this.

10. See, for example, Noble (1984) and (1977); Winner (1992); and Goldman (1992).

References

Bijker, Wiebe. 1995. *Of Bicycles, Bakelites, and Bulbs: Toward a Theory of Sociotechnical Change*. Cambridge, MA: MIT Press.

Brunner, John. 1975. *Shockwave Rider*. New York: Ballantine.

Davis, Stan, and Meyer, Christopher. 1998. *Blur: The Speed of Change in the Connected Economy*. Reading, MA: Addison Wesley.

Dell, Michael, and Fredman, Catherine. 1999. *Direct from Dell: Strategies that Revolutionized an Industry*. New York: Harper.

Derber, Charles. 1998. *Corporation Nation: How Corporations Are Taking Over Our Lives and What We Can Do About It*. New York: St. Martin's Press.

Drexler, K. Eric. 1992. *Nanosystems: Molecular Machinery, Manufacturing, and Computation*. New York: Wiley.

Drucker, Peter. 1946. *The Concept of the Corporation*. New York: John Day.

Ellul, Jacques. 1954. *La Technique ou L'enjeu du siècle*. Paris: A. Colin. (English trans. *The Technological Society*. New York: Knopf, 1964).

Fudpucker, Wilhelm E. 1978. *Shepherd of Becoming*. London: Noosphere.

Gibson, William. 1987. *Count Zero*. New York: Ace.

——. 1984. *Neuromancer*. New York: Ace.

Gilder, George. 1989. *Microcosm: The Quantum Revolution in Economics and Technology*. New York: Simon and Schuster.

Goldman, Steven L. 1992. "No Innovation Without Representation: Technological Action in a Democratic Society." Pp. 148–60 in Stephen H. Cutcliffe, Steven L. Goldman, Manuel Medina, and José Sanmartín, eds., *New Worlds, New Technologies, New Issues*. Bethlehem, PA: Lehigh University Press.

Goldman, Steven L., and Preiss, Kenneth. 1991. *21st Century Manufacturing Enterprise Strategy: An Industry-Led View*. Bethlehem, PA: Iacocca Institute.

Goldman, Steven L., Preiss, Kenneth, and Nagel, Roger N. 1995. *Agile Competitors and Virtual Organizations: Strategies for Enriching the Customer*. New York: Van Nostrand Reinhold.

———. 1996. *Cooperate to Compete: Building Agile Business Relationships.* New York: Van Nostrand Reinhold.

Jacques, Elliot. 1991. *The Changing Culture of a Factory.* New York: Dryden Press.

Halberstam, David. 1986. *The Reckoning.* New York: Morrow.

Harrison, Bennett. 1994. *Lean and Mean: The Changing Landscape of Corporate Power in the Age of Flexibility.* New York: Basic Books.

Kanigel, Robert. 1997. *The One Best Way: Frederick Winslow Taylor and the Enigma of Efficiency.* New York: Viking.

Mitcham, Carl, and Grote, Jim, eds. 1984. *Theology and Technology: Essays in Christian Analysis and Exegesis.* Lanham, MD: University Press of America.

Moravec, Hans P. 1999. *Robot: Mere Machine to Transcendent Mind.* New York: Oxford University Press.

Noble, David. 1977. *America by Design: Science, Technology, and the Rise of Corporate Capitalism.* New York: Knopf.

———. 1984. *Forces of Production: A Social History of Industrial Automation.* New York: Knopf.

Stalk, George, and Hout, Thomas M. 1990. *Competing Against Time: How Time-Based Competition is Reshaping Global Markets.* New York: Free Press.

Stephenson, Neil. 1992. *Snow Crash.* New York: Bantam.

———. 1996. *The Diamond Age.* New York: Bantam.

Warnecke, Hans-Jürgen. 1993. *The Fractal Company: A Revolution in Corporate Culture.* Trans. Maurice Claypole. New York: Springer-Verlag.

Winner, Langdon, ed. 1992. *Democracy in a Technological Society.* Boston, MA: Kluwer.

Womack, James P., and Jones, Daniel T. 1996. *Lean Thinking: Banish Waste and Create Wealth in Your Corporation.* New York: Simon and Schuster.

Womack, James P., Jones, Daniel T., and Roos, Daniel. 1991. *The Machine That Changed the World: The Story of Lean Production.* New York: Harper.

Bibliography

As a continuously developing field, science, technology, and society studies is blessed with a wealth of recent literature. What follows is thus a very selective introduction to a diverse body of work. The emphasis is on currently available English-language books that address STS as a whole rather than particular issues within the field. Reference is also provided to a few important volumes that are no longer in print, with some modest highlighting of otherwise neglected approaches that we think ought to be incorporated into STS work. For more bibliographic references to the STS literature, see the references to each chapter in the present book, and for a more comprehensive list, see Cutcliffe (2000).

Biagioli, Mario, ed. *The Science Studies Reader*. New York: Routledge, 1999. Pp. xviii, 590. Thirty-six essays mostly written in the 1990s, in which the authors take a sociocultural approach and focus on the physical and biological sciences. Includes discussion of the dichotomy between realism and constructionism; comparison of cognitive styles across different cultures, time periods, and disciplines; gender issues; scientific practice; scientific authorship and credit; and the ways non-Western cultures understand and conceptualize nature. Includes two bibliographies.

Bijker, Wiebe. *Of Bicycles, Bakelites, and Bulbs: Toward a Theory of Sociotechnical Change*. Cambridge, MA: MIT Press, 1995. Pp. x, 380. Collects three seminal studies by a leading practitioner of the Dutch school of social constructivism.

Bijker, Wiebe E., Hughes, Thomas P., and Pinch, Trevor J., eds. *The Social Construction of Technological Systems: New Directions in the Sociology and History of Technology*. Cambridge, MA: MIT Press, 1987. Pp. x, 405. This collection of thirteen studies emerged from a 1984 workshop held at the University of Twente, The Netherlands. It may be said to have opened up the social constructivist approach with regard to technology.

Bridgstock, Martin, Burch, David, Forge, John, Laurent, John, and Lowe, Ian. *Science, Technology and Society: An Introduction*. New York: Cambridge University Press, 1998. Pp. xii, 276. Australian textbook offering a comprehensive introduction to the human, social, and economic aspects of science and technology. Authors are all in the Faculty of Science and Technology at Griffith University, Queensland.

Chalk, Rosemary, ed. *Science, Technology, and Society: Emerging Relationships, Papers from "Science," 1949–1988*. Washington, DC: American Association for the Advancement of Science, 1988. Pp. vi, 262. Collects over eighty representative articles under ten headings: science and responsibility, science and freedom, science and ethics, the human side of science, scientists and citizens, science and the modern world, fraud and misconduct in science, professional rights and duties in the health sciences, science and risk, and science and national secruity. Reveals a consistent and intensive involvement of the scientific community in STS issues.

Collingridge, David. *The Social Control of Technology*. New York: St. Martin's Press, 1982. Pp. 200. Clear statement of what has become known as the "Collingridge dilemma": early in the development of a technology, when it would be easy to change, we don't know enough to want to change it; later, when we have enough knowledge to want to change it, it is extremely difficult to change.

Collins, Harry, and Pinch, Trevor. *The Golem: What You Should Know About Science*. 2nd ed. New York: Cambridge University Press, 1998. Pp. xix, 192. Presents a view of science as "fallible and untidy" through a series of seven sociologically oriented case studies focusing on issues in science at their "controversial" stage before they reach stability or general agreement regarding what counts as acceptable scientific knowledge and understanding. Primarily intended for the student and citizen who wishes to take part in the democratic process of a technological society.

———. *The Golem at Large: What You Should Know About Technology*. New York: Cambridge University Press, 1998. Pp. xi, 163. Seven short case studies promoting the thesis that technology "is not an evil creature but it is a little daft" and that its problems are due more to our clumsy learning processes than anything else. Does for technology what the editors' earlier collection did for science.

Cozzens, Susan E., and Gieryn, Thomas F., eds. *Theories of Science in Society*. Bloomington, IN: Indiana University Press, 1990. Pp. 264. A collection of ten original papers that offer a range of theoretically informed case studies focused on understanding science in its societal context, rather than as an activity separate from human roles and functions.

Cutcliffe, Stephen H. *Ideas, Machines, and Values: An Introduction to Science, Technology, and Society Studies*. Lanham, MD: Rowman and Littlefield, 2000. An introduction to STS which provides a historical review of the field's evolution over three decades, an overview of the major disciplinary and interdisciplinary approaches entailed, a summary of its institutional organization, and suggestions for future directions. Includes two helpful bibliographies.

Durbin, Paul T. *Social Responsibility in Science, Technology, and Medicine*. Bethlehem, PA: Lehigh University Press, 1992. Pp. 230. A call for socially responsible, public interest activism in identifying and controlling the problematic aspects of science, technology, and medicine by a leading philosopher of technology. Specific areas addressed include science education, health care, the media and politics, biotechnology, computers, nuclear weapons and power, and the environment.

Fuglsang, Lars. *Technology and New Institutions: A Comparison of Strategic Choices and Technology Studies in the United States, Denmark, and Sweden*. Copenhagen: Academic Press, 1993. Pp. 226. A good comparative study of the state of the field of STS studies in early 1990s.

Fuller, Steve. *Philosophy, Rhetoric, and the End of Knowledge: The Coming of Science and Technology Studies*. Madison, WI: University of Wisconsin Press, 1993. Pp. xxii, 421. Argues that STS and science and technology studies grew out of nineteenth-century studies in the history and philosophy of science. Not an introductory text, but written for those who seek greater depth in the background of STS.

Guston, David H., and Keniston, Kenneth, eds. *The Fragile Contract: University Science and the Federal Government*. Cambridge, MA: MIT Press, 1994. Pp. xiv, 244. Eleven papers growing out of a 1991–1992 workshop. The "Introduction: The Social Contract for Science," is one of the best brief overviews available on contemporary U.S. science policy.

Hess, David. *Science Studies: An Advanced Introduction*. New York: New York University Press, 1997. Pp. vii, 197. An overview of key concepts promoting an integrated framework. Includes recent developments in philosophy, sociology, anthropology, history, cultural studies, and feminist studies.

Jasanoff, Sheila, Markle, Gerald E., Petersen, James C., and Pinch, Trevor J., eds. *Handbook of Science and Technology Studies.* Thousand Oaks, CA: Sage, 1995. Pp. xv, 820. A controversial but important compilation of twenty-eight specially commissioned articles intended to provide state-of-the-art summaries of key areas and issues within the field. Includes a lengthy bibliography.

Latour, Bruno. *Science in Action: How to Follow Scientists and Engineers through Society.* Cambridge, MA: Harvard University Press, 1987. Pp. viii, 274. Latour opens up the "black box" of modern "techno-science" to more fully understand what goes on in the process of creating knowledge and artifacts by following scientists "inside" their laboratories and then "outside" along their associational networks. He finds that the sociological dimension is an elementary and integral component of technoscience, not separate from it. Thus, both technological and social deterministic approaches are misguided.

———. *Pandora's Hope: Essays on the Reality of Science Studies.* Cambridge, MA: Harvard University Press, 1999. Pp. x, 324. Latour responds to critics who have accused him of denying the existence of an independent reality by substituting a notion of a "collective" of "circulating references" between human and nonhumans for the "modernist" subject-object dichotomy that separates epistemology, nature, politics, mind, and society as though they were all independent elements. This alternative is the "hope" at the bottom of Pandora's box. An important statement of Latour's evolving views.

Longino, Helen. *Science as Social Knowledge: Values and Objectivity.* Princeton: Princeton University Press, 1990. Pp. xi, 262. Draws on feminist theory in the areas of human biological evolution to argue for a "contextual empiricism" that reconciles scientific objectivity and the shaping influence of social values.

Lynch, Michael. *Scientific Practice and Ordinary Action: Ethnomethodology and Social Studies of Science.* New York: Cambridge University Press, 1994. Pp. 354. Criticism of developments in science studies that have led to the thesis that scientific knowledge is socially constructed.

MacKenzie, Donald, and Wajcman, Judy, eds. *The Social Shaping of Technology.* 2nd ed. Philadelphia: Open University Press, 1999. Pp. xvii, 462. Thirty representative articles on four themes: general issues, technological production, reproductive technology, and military technology.

McGinn, Robert E. *Science, Technology, and Society.* Englewood Cliffs, NJ: Prentice Hall, 1991. Pp. xx, 304. An introductory textbook which is nevertheless crammed with information and sophisticated interpretations of STS relationships. Organized into three parts. Part one develops foundational materials useful for analyzing science and technology in society. Part two explores the influence of science and technology on modern society. Part three considers the influence of modern society on science and technology.

Mowery, David C., and Rosenberg, Nathan. *Paths of Innovation: Technological Change in 20th-Century America*. Cambridge: Cambridge University Press, 1998. Pp. x, 214. An examination of U.S. innovation that seeks to explore the relationship between technology and economic growth with an eye toward influencing policy-makers regarding the importance of investment in technological development. Three areas of research-intensive technology have dominated America in this century–internal combustion engines, chemicals, and electricity and electronics.

Pacey, Arnold. *Meaning in Technology*. Cambridge: MIT Press, 1999. Pp. viii, 264. The most recent of a series of excellent books by Pacey on the interaction of technology and society over the past thousand years. This volume explores the individual's experience with technology from the engineer to the worker to the consumer. He also examines technology with regard to our sense of place, the environment, gender, and creativity. Concludes with a call for a more "people-centered" technology. See also his earlier works, *The Maze of Ingenuity*, 2nd ed. (MIT Press, 1992), and *The Culture of Technology* (MIT Press, 1993).

Sarewitz, Daniel. *Frontiers of Illusion: Science, Technology, and the Politics of Progress*. Philadelphia: Temple University Press, 1996. Pp. xi, 235. Critiques five myths about science: the myths of infinite benefit, of unfettered research, of accountability, of authoritativeness, and of the endless frontier. By a geoscientist realist who rejects social constructivism but nonetheless recognizes the significance of the broader societal context in which science is embedded.

Simon, Julian L., with Beisner, E. Calvin, and Phelps, John, eds. *The State of Humanity*. Cambridge, MA: Blackwell, 1995. Pp. x, 694. Fifty-eight original articles arguing an optimistic interpretation of the impact of science and technology on society.

Teich, Albert H. *Technology and the Future*. 8th ed. New York: St. Martin's Press, 2000. Pp. xix, 341. Twenty-seven representative articles. Since the 1st ed. under the title *Technology and Man's Future* (1972), this has served as a standard anthology for many introductory STS courses.

Volti, Rudi. *Society and Technological Change*. 3rd edition. New York: St. Martin's Press, 1995. Pp. xiv, 315. A textbook review of perspectives, theories, and facts on the consequences of technological

change and the forces that produced it. Includes a general introduction on the nature of technology, an examination of how new
technologies originate and diffuse, illustrations of the interactions
between technological and social change in health and medicine,
the environment, work, communication, and warfare, and an
analysis of the control of technology.

———, ed. *The Facts on File Encyclopedia of Science, Technology, and Society*, 1999. 3 vols. Pp. ix, 1201. Approximately 900 individual
entries from abacus to zipper that provide the societal context in
which science and technology have emerged, been developed, and
put to use. Also included are entries that analyze the philosophical and methodological underpinnings of science and technology.
Includes cross-references to related topics, suggestions to further
readings where available, and an overview bibliography.

Wajcman, Judy. *Feminism Confronts Technology*. University Park, PA:
Penn State University Press, 1991. Pp. x, 184. Analysis of the relationship between gender and technology from an antiessentialist,
social constructivist perspective. Areas of emphasis include
women and work, reproductive technology, domestic household
technology, and the built environment.

Webster, Andrew. *Science, Technology, and Society*. Rutgers University
Press, 1991. Pp. viii, 181. An introduction to the sociological analysis of science and technology, broadly construed, by a professor of
sociology at Anglia College, Cambridge, UK. Includes discussion of
biotechnology and genetic engineering, technology transfer, feminist and radical critiques of science.

Westrum, Ron. *Technologies & Society: The Shaping of People and Things*.
Belmont, CA: Wadsworth Publishing Co., 1991. Pp. xx, 394. A sociologically oriented introductory textbook that focuses on the ways
technology and people "intertwine" with each other. Westrum is
particularly interested in questions of how technology is created in
the manner that it is, as well as society's attempts to control its
effects. Contains two especially useful chapters on the idea of technology "sponsorship." Although currently out of print, the volume
is available from the author, Dept. of Interdisciplinary Technology,
Eastern Michigan University, Ypsilanti, MI 48197.

Winner, Langdon. *The Whale and the Reactor: A Search for Limits in an
Age of High Technology*. Chicago: University of Chicago Press,
1987. Pp. xiv, 200. A thoughtful and thought provoking collection
of integrated essays by one of the leading STS critics of modern
technoscientific culture. Among the key ideas presented are the
notion that "technologies are forms of life" and that "artifacts have
politics." A call to technological sleepwalkers to awaken.

Yager, Robert E., ed. *Science/Technology/Society as Reform in Science Education*. Albany, NY: State University of New York Press, 1996. Pp. xi, 339. Collects twenty-nine papers on five themes: STS as a reform of science education, what the STS approach can accomplish, what the STS approach demands, STS science education initiatives outside the United States, and STS science education supporting activities.

Ziman, John. *An Introduction to Science Studies: The Philosophical and Social Aspects of Science and Technology*. New York: Cambridge University Press, 1984. Pp. xi, 203. "The conventional description of the scientific community as a republic or oligarchy of autonomous scientists, exchanging communications for personal recognition is not yet out of date, but it must be radically modified to take full account of the structures that have grown up to coordinate and *manage* scientific work, even in its 'purest' and most academic modes" (p. 139). Balanced review of the basic issues by a respected physicist who has also been a leader of STS studies in Great Britain. His *Teaching and Learning about Science and Society* (New York: Cambridge University Press, 1980) is also important.

———. *Prometheus Bound: Science in a Dynamic Steady State*. New York: Cambridge University Press, 1994. Pp. 289. On science policy and the reorganization of scientific research in the face of a leveling off of expenditures. Calls for social *space* for personal initiative and creativity; *time* for ideas to grow to maturity; *openness* to debate and criticism; hospitality toward *novelty*; and respect for specialized *expertise*.

Leading STS Journals

A large number of periodicals include articles relevant to the STS field. Examples include journals in the history of science (such as *Isis*), the history of technology (such as *Technology and Culture*), public policy (*Minerva* and *Public Understanding of Science*), applied philosophy (from *Applied Philosophy* to the *Journal of Business Ethics*, *Science and Engineering Ethics*, and *Ethics and Information Technology*), and popular culture (such as *American Heritage of Invention and Technology* and the *Journal of Popular Culture*). Newsletters such as the *Science, Technology & Society Curriculum Newsletter* (Lehigh University), the *Ethical and Policy Issues Perspectives on the Professions* (Illinois Institute of Technology), and the *Professional Ethics Report* (American Association for the Advancement of Science) are also quite important. The following nevertheless lists only the major journals publishing in broad areas of the STS field.

Bulletin of Science, Technology, and Society (1981–present). Bi-monthly. Founded by Rustum Roy, Penn State University. Publication of the National Association for Science, Technology, and Society (NASTS). Oriented toward STS curriculum development and the impacts of science and technology on society.

IEEE Technology and Society Magazine (1982–present). Quarterly. Publication of the Society for the Social Implications of Technology of the Institute of Electrical and Electronic Engineers. Emphasizes the work of engineers reflecting on the social context and implications of their work.

Issues in Science and Technology (1984–present). Quarterly. Publication of the National Academies of Science and of Engineering and by the Cecil and Ida Green Center for the Study of Science and Society, University of Texas, Dallas. Articles tend toward public policy, industry, and R&D related issues.

Perspectives on Science (1993–present). Quarterly. Publishes articles offering perspectives from history, philosophy, and sociology of science.

Science, Technology, & Human Values (1975–present). Quarterly. Journal of the Society for Social Studies of Science. Continues "Newsletter on Science, Technology, and Human Values," which was published by the Harvard University Program on Science, Technology, and Public Policy. The most broad-based scholarly journal placing some emphasis on social science analyses of science and technology.

Social Studies of Science (1971–present). Quarterly. Focuses on the social construction of science and technology.

Technology in Society (1979–present). Quarterly. Emphasizes issues of the management and administration of technoscience.

Index

This index covers the introduction and ten essays exclusive of the reference lists, with a focus on proper names. An "n" attached to a page number indicates a note (with "nn" indicating a plural), and is followed by the numbers of the relevant note (or notes), thus providing more specific locators than simple page references for many of the entries. The letter "h" attached to a number indictes a headnote. The pages for the authors of essays are indicated by the word "author."

SUNY series in Science, Technology, and Society
Sal Restivo and Jennifer L. Croissant, editors

Peerless Science: Peer Review and U.S. Science Policy, Daryl E. Chubin and Edward J. Hackett

Scientific Knowledge in Controversy: The Social Dynamics of the Fluoridation Debate, Brian Martin

Social Control and Multiple Discovery in Science: The Opiate Receptor Case, Susan E. Cozzens

Human Posture: The Nature of Inquiry, John A. Schumacher

Women in Engineering: Gender, Power, and Workplace Culture, Judith S. McIlwee and J. Gregg Robinson

The Professional Quest for Truth: A Social Theory of Science and Knowledge, Stephan Fuchs

The Value of Convenience: A Genealogy of Technical Culture, Thomas F. Tierney

Math Worlds: Philosophical and Social Studies of Mathematics and Mathematics Education, Sal Restivo, Jean Paul Van Bendegem, Roland Fischer, eds.

Knowledge Without Expertise: On the Status of Scientists, Raphael Sassower

Controversial Science: From Content to Contention, Thomas Brante, Steve Fuller, William Lynch, eds.

Science, Paradox, and the Moebius Principle: The Evolution of a "Transcultural" Approach to Wholeness, Steven M. Rosen

In Measure, Number, and Weight: Studies in Mathematics and Culture, Jens Høyrup

On the Shoulders of Merchants: Exchange and the Mathematical Conception of Nature in Early Modern Europe, Richard W. Hadden

Ecologies of Knowedge: Work and Politics in Science and Technology, Susan Leigh Star

Energy Possibilities: Rethinking Alternatives and the Choice-Making Process, Jesse S. Tatum

Science without Myth: On Constructions, Reality, and Social Knowledge, Sergio Sismondo

The Science of Empire: Scientific Knowledge, Civilization, and Colonial Rule in India, Zaheer Baber

In and About the World: Philosophical Studies of Science and Technology, Hans Radder

Social Constructivism as a Philosophy of Mathematics, Paul Ernest

Beyond the Science Wars: The Missing Discourse about Science and Society, Ullica Segerstråle, ed.

Science, Technology, and Democracy, Daniel Lee Kleinman, ed.